Mobile Application Development
with SMS and the SIM Toolkit

Scott B. Guthery

Mary J. Cronin

McGraw-Hill

New York • Chicago • San Francisco • Lisbon
London • Madrid • Mexico City • Milan • New Delhi
San Juan • Seoul • Singapore • Sydney • Toronto

McGraw-Hill
A Division of The McGraw-Hill Companies

Copyright © 2002 by McGraw-Hill Companies, Inc. All rights reserved. Printed in the United States of America. Except as permitted under the United States Copyright Act of 1976, no part of this publication may be reproduced or distributed in any form or by any means, or stored in a data base or retrieval system, without the prior written permission of the publisher.

1 2 3 4 5 6 7 8 9 0   DOC/DOC   0 9 8 7 6 5 4 3 2 1

ISBN 0-07-137540-6

The sponsoring editor for this book was Marjorie Spencer, the editing supervisor was Steven Melvin, and the production supervisor was Sherri Souffrance. It was set in Vendome by Patricia Wallenburg.

Printed and bound by R. R. Donnelley & Sons Company.

McGraw-Hill books are available at special quantity discounts to use as premiums and sales promotions, or for use in corporate training programs. For more information, please write to the Director of Special Sales, Professional Publishing, McGraw-Hill, Two Penn Plaza, New York, NY 10121-2298. Or contact your local bookstore.

Throughout this book, trademarked names are used. Rather than put a trademark symbol after every occurrence of a trademarked name, we use names in an editorial fashion only, and to the benefit of the trademark owner, with no intention of infringement of the trademark. Where such designations appear in this book, they have been printed with initial caps. The 3GPP TS 31.102 Third Generation Mobile System Release 1999, v.3.2.0 is the property of ARIB, CWTS, ETSI, T1, TTA and TTC who jointly own the copyright in it. It is subject to further modifications and is therefore provided to you "as is" for information purpose only. Further use is strictly prohibited.

Information contained in this book has been obtained by The McGraw-Hill Companies, Inc., ("McGraw-Hill") from sources believed to be reliable. However, neither McGraw-Hill nor its authors guarantee the accuracy or completeness of any information published herein, and neither McGraw-Hill nor its authors shall be responsible for any errors, omissions, or damages arising out of use of this information. This work is published with the understanding that McGraw-Hill and its authors are supplying information, but are not attempting to render engineering or other professional services. If such services are required, the assistance of an appropriate professional should be sought.

 This book is printed on recycled, acid-free paper containing a minimum of 50 percent recycled, de-inked fiber.

This book is dedicated to

Tyler Guthery
Rebecca Cronin
Johanna Cronin

Our next generation

CONTENTS

Foreword	xi
Acknowledgments	xiii

1 Introducing SMS and the SIM — 1

Foundations and Definitions	4
SMS and SIM in the Network Context	7
Protocol Stacks	9
The Role of Standards	11
Preview of Coming Chapters	16
Summary	16

2 Basic SMS Messaging — 19

Connecting the Handset	20
Communicating with the Handset	21
Communicating with the Network	24
Hello, Mobile World	25
Summary	38

3 Details of SMS-SUBMIT and SMS-DELIVER — 39

Numbering Plans and Mobile Telephone Numbers	42
SMS_SUBMIT	42
Protocol Identifier	47
Data Coding Scheme	49
Concatenated Short Messages	51
"You've Got Mail"	52
Application Port Addressing	53
SIM Toolkit Security	54
Enhanced Messaging Services	54
Sounds, Pictures, and Animations	56
Internet E-Mail	60
SMS_DELIVER	61
Summary	63

4	**SMS Integration**	**65**
	Summary	78
5	**SMS Brokers**	**79**
	Summary	92
6	**SMS in an Airport Logistics Application**	**95**
	SMS Case Study: Atraxis	96
	Project Background	97
	Focus on the Essentials	98
	Design and Development Process	99
	The Action on the Ground	101
	Project Performance Review	103
	Evaluating the Business Results	104
	Summary	105
7	**The SIM**	**107**
	Smart Cards 101	111
	The Evolution of the SIM	115
	Who Are You?	118
	Evolution of SIM Standards	119
	The Birth of the SIM Application Toolkit	122
	The SAT API	127
	The USAT Interpreter	128
	Summary	130
8	**SIM Toolkit API: Proactive Commands and Event Download**	**131**
	Proactive Commands	133
	Details of SIM Toolkit Commands	142
	Application Commands	143
	Smart-Card Proactive Commands	146
	General Purpose Communication Commands	146
	System Commands	147

Contents

	Event Download	148
	Summary	155
9	**End-to-End Security for SMS Messages**	**157**
	Security Parameter Indicator (SPI)	161
	Ciphering Key Identifier (KIc) and the Key Identifier (KID)	162
	Toolkit Application Reference (TAR)	164
	Counter (CNTR)	165
	Padding Counter (PCNTR)	165
	Redundancy Check (RC), Cryptographic Checksum (CC), or Digital Signature (DS)	166
	Secured SMS Message Example	166
	Proof of Receipt	168
	Pairing a Sent Message with its Response	170
	Summary	172
10	**The SmartTrust Microbrowser and the 3GPP USAT Interpreter**	**173**
	Some More SIM Toolkit History	174
	A Short History of Byte Code Interpreters on Smart Cards	176
	Sonera SmartTrust WIB	180
	The 3GPP USAT Interpreter	188
	Remote Procedure Call Using the USAT Interpreter	193
	Summary	195
11	**The USAT Interpreter at Work**	**197**
	Business Drivers	198
	Technology Overview	200
	Starting With SMS	200
	From WAP to One Integrated Portal	202
	Integrating with the Microbrowser	204
	Moving to Mobile Banking and M-Commerce	204
	From the User Point of View	205
	Implementation Challenges and Strategies	207
	Bottom-Line Benefits	209
	Lessons Learned	210

12 The USAT Virtual Machine and SIM Toolkit Programs — 211

- Variants of the USAT Virtual Machine — 214
- Virtual Machine Architectures — 216
- The USAT Virtual Machine from Microsoft — 218
- Real-Time Travel Example — 224
  - Central versus Local Storage of Personal Information — 224
- Java Card™ SIMs — 235
- Installation of USAT Virtual Machine Programs — 235
- Summary — 237

13 Smart Signatures for Secure Mobile Commerce — 239

- Starting With the Mobile Customer — 241
- SmartSignature Features — 243
  - Forms and Templates — 243
  - Keys and PINs — 244
  - Menu Design — 244
  - Changing Service Providers — 245
- Mobile Certification and Trust Using SmartSignature — 248
  - Trust Relationships for Making the Transaction — 251
  - Trust Relationship for Enabling the Transaction — 252
  - Certification Authorities — 253
- Business Enablers of SmartSignature — 253
  - SmartSignature in Operation — 254
  - SmartSignature in the Setup Phase — 256
- Managing a Large Pilot of SmartSignature — 258
  - Pilot Background — 258
  - The Key Participants — 259
  - Revenue Model — 260
  - Pricing of SmartTrust Components — 260
  - Security in a Mobile Trust Hierarchy — 261
  - Lessons of the Pilot Delivery — 262
  - Importance of the Customer's Experiences — 262
  - Implications to the Business Model — 263
  - Implications for SmartTrust Business Strategy — 263
- Next Steps with SmartSignature — 264

14 The ETSI Smart Card Platform — 267

Managed Data Sharing Using Access Control Lists — 269
Associating Access Control Lists with Files — 272
Coding Access Control Rules — 274
Access Mode TLV — 275
Key References — 276
Boolean Expressions of Key References — 278
Key Reference Semantics — 280
Authentication of Key References — 283
Application Activation and Concurrent Execution — 284
    The Application Directory and
        Application Activation — 285
Application Activation and Concurrent Execution — 285
    Application Selection — 287
    Concurrent Application Execution — 288
Summary — 289

APPENDIX Standards for SMS and the SIM — 291

Third Generation Partnership Project (3GPP) — 291
    3GPP Technical Specification Group T
        (Terminals)—Working Group 2 Mobile
        Terminal Services and Capabilities — 291
    3GPP Technical Specification Group T
        (Terminals)—Working Group 3 Universal
        Subscriber Identity Module (USIM) — 292
European Telecommunications Standards
    Institute (ETSI) Smart Card Project — 293
International Organization for Standardization (ISO) — 294

Index — 295

FOREWORD

The success story of GSM is also the success story of the SIM. Every subscriber needs a SIM and there is no service without it. This is unlike some other systems where the micro-computer in the smart card offers just an additional service which may or may not be used by the customer. With more than 600 million subscribers worldwide, GSM is by far the largest application employing smart cards and it has taken the smart card industry from its infancy to adulthood. GSM is closely linked with the introduction of mass production of smart cards and the ever increasing requirements of the SIM have given a huge impetus not only to the technological advancement of the microcomputer itself, be it the memory provided by today's chips or their electrical parameters, but also to the development of operating systems, application provision and programming interfaces of smart cards in general.

Only in the last few years has the telecommunications community at large begun to recognize the importance of the contribution of the SIM to the success of GSM. At the birth of GSM, the goal of the SIM was to provide an unprecedented level of security in mobile communications. The SIM also "freed" the mobile phone from the subscription and security aspects. This created, for the first time, a virtually global terminal market.

Today, the SIM offers more than just these two things. The standardization of the SIM Application Toolkit and now the Interpreter, together with the advancement in the hardware platform for the SIM created an ever advancing platform for *secure* value added services at the discretion and under the control of the operator and the service provider. Content is the magic word and it will even be more so in the future.

This book is the first comprehensive presentation of the technical issues, including a very detailed introduction to SMS, which currently form the basis of Toolkit and Interpreter. It combines these technical details with thorough presentations of life-examples, making it also a useful source for marketing people with a technical background. This is what Toolkit and Interpreter need: more marketing attention in the higher ranks of the operators and service providers. Everybody there knows WAP but who has heard of Toolkit and Interpreter, let alone how to make money by deploying them in an innovative manner? WAP-like handset-based services and SIM-based Toolkit and Interpreter services do not exclude each other, they can complement each other in an optimal way.

The fact that this book exists at all, illustrates one of the benefits of having a single standard over multiple proprietary solutions. Toolkit and Interpreter have been standardized for SIM and USIM by ETSI and the 3GPP. They are solution based standards. The history of GSM has clearly shown that only solution based standards can provide the high level of interoperability between system components necessary for a multivendor environment and the independence from disparate proprietary solutions which are essential for the long-term success. I hope and expect this book to spread the knowledge of these great tools and thus to broaden the penetration of the SIM as a platform for value added services providing content.

I also expect this book to cause a lot of interesting and, I am sure, controversial discussions on technical and market aspects of Toolkit and Interpreter as well as on some of the "historical" statements. Having been involved in the standardization of the SIM from its beginning and believing in its future as being more than a security device, I am looking forward to these discussions. They will certainly give a new impetus to the world of the UICC as *the* smart card platform for (mobile) communications.

Dr. Klaus Vedder
Giesecke & Devrient
Chairman ETSI EP SCP (Smart Card platform)
Chairman 3GPP TSG-T3 (USIM)
email: klaus.vedder@gdm.de

ACKNOWLEDGMENTS

The development of international SMS and SIM standards and interoperable application platforms for SIM and SMS requires a collective effort that spans many countries and points of view. So it's no surprise that this book draws heavily on the expertise and experience of many, many participants in the standards development process. We owe a large debt of gratitude to all the busy people who read early versions of chapters, answered complicated questions promptly, and generously shared their recollections and documentation of the early decisions that helped to shape today's SMS and SIM standards and point the way to the next generation applications. We have named many of these below, but fully realize that the list is by no means complete—so thank you to all the colleagues in 3GPP Terminals (T) and ETSI Smart Card Platform (SCP) standards bodies whose standards work literally made this book possible and to the denizens of various newsgroups and listserv lists including alt.technology.smartcards and eurowireless.

Likewise, the case studies that illustrate how operators and corporations are using SIM and SMS applications exist primarily because of the generosity and responsiveness of managers and practitioners who devoted many hours to answering questions, supplying data and detailed explanations, and carefully reviewing early drafts of the cases. Special thanks to Anselmo A. Mazzoleni of the Atraxis Group in Zurich and to Paul Aebi of Swisscom Ltd for their help in completing the Atraxis case write up, to Thomas Bruun Pedersen of Sonofon in Denmark for the extensive interviews and follow up on the Sonofon case, and to Jarkko Rossi, Lars-Erik Sellin, and Werner Freystätter of SmartTrust for their insights and explanations about the technical and business complexities of security for mobile commerce and for multiple updates and reading of drafts. Also thanks to Ari-Pekka Kitinoja of Sonera and Jouni Heinonen of Setec for essential background details and explanation. Our gratitude also goes to Anders Sellin of SmartTrust for his essential early help in framing case topics and introducing us to case prospects among his many contacts in the SIM applications world.

Once the book reached its final draft, three experts took the time to read the entire manuscript closely and make valuable comments and corrections. Our appreciation to Nigel Barnes, Jean-Francois Rubon, and Kristian Woodsend for this invaluable service.

Throughout the research and writing process, we called on a number of colleagues to supply background information and help clarify specific points of standards and application implementation. Among the many who responded to these queries, special thanks to David Birch, Peter De Vijt, Bertrand du Castel, David Everett, Tony Guilfoyle, Colin Hamling, Mark Kamers, Roger Kehr, Tim Jurgensen, Hans-Joachim Knobloch, Michael Meyer, Pierre Paradinas, David Pecham, Patrice Peyret, Jochaim Posegga, Fred Renner, Edouard Richard, Wolfgang Salge, Lars-Erik Sellin, Gerry Smith, Jean-Jacques Vandewalle, John Wood and last but definitely not least, Klaus Vedder.

The tables and graphics that are reprinted herein with permission of ETSI, Atraxis, Setec, SmartTrust, and Sonofon enhance the readability of the book, and we gratefully acknowledge their help.

A heartfelt salute to those closer to home who supported our research, writing, and updating efforts throughout the whole process. To the entire staff of Mobile-Mind, and in particular to Dan Eichenwald, Peter Laing, Scott Marks, Scott Olihovik and Perry Spero, we are happy to tell the world that we couldn't have made it to the last page without your day-to-day contributions. A sincere thank you to Marjorie Spencer, our excellent and very patient editor, and to Rob Robertson, our agent, for his confidence that this book was meant to be.

Finally, we fully recognize that even with the best of support and expert advice, in the fast-changing world of SMS and SIM applications there are bound to be changes and inaccuracies in any description that becomes frozen in print. We hope that readers will send us their comments and corrections to help improve the next edition.

Scott B. Guthery
sguthery@mobile-mind.com

Mary J. Cronin
mcronin@mobile-mind.com

CHAPTER 1

Introducing SMS and the SIM

Wireless devices have overtaken every other technology—including the Internet—in global adoption. By 2003 more than a billion people will be using a wireless phone or personal digital assistant (PDA) for voice and data communications. Three factors that have helped to drive this phenomenal growth have also inspired this book:

1. The worldwide availability and popularity of an inexpensive Short Message Service (SMS);
2. The evolution of the Subscriber Identity Module (SIM) inside GSM phones into a standardized and secure application platform for GSM and next-generation networks; and
3. The demand for applications that let people use their mobile phones for more than just talking.

Let's take a quick look at how SMS and the SIM have contributed to the growth of wireless applications and then discuss what you can expect to learn from this book.

The number of SMS messages sent every month has risen from about 1 billion messages in July 1999 to more than 20 billion in July 2001, with projections that the total number of SMS messages exchanged in 2001 will top 200 billion. These SMS exchanges range from simple text greetings or questions sent between individual subscribers (sometimes called "texting") to news and information services offered by the wireless carriers, to more advanced applications offered by third parties such as retrieving data from a corporate sales database or mobile banking. One result of all this texting and other SMS activity is that wireless carriers now view SMS as an important source of revenues. Another outcome is that hundreds of millions of subscribers are ready and eager to try out interesting new services based on SMS. But to move beyond the basic text message delivery and create applications that can be customized and trusted, developers need a standardized and secure application platform. That's where the SIM comes in.

The SIM is a smart chip that was designed as a secure, tamper-resistant environment for the cryptographic keys that GSM carriers use to authenticate individual subscribers to the network connection and track those subscribers' activities once they are on the air. The SIM maintains a constant connection to the network as long as the mobile device remains on. This location-aware, authenticated connection is what allows subscribers to "roam" from network to network around the world and, very importantly from the viewpoint of the carrier, the SIM keeps track of and reports on the subscriber's network usage and roaming activity so that the carrier can bill customers accurately.

Introducing SMS and the SIM

The only way to ensure that the SIM can accomplish its handoff of subscribers from one network to another without interrupting communication is to base all of its functions on very detailed international standards. Every GSM equipment manufacturer and carrier adheres to these standards, which cover everything from the physical size and characteristics of the chip to the way it handles and stores incoming information. Anyone developing applications that interact with the SIM also has to become familiar with the relevant standards and keep up with changes. This book describes the most important standards in detail and points readers to online sources of complete standard documentation and updates.

The SIM is also an essential part of the move to higher speed and more capable "next-generation" wireless networks, discussed later in this chapter. Because the 2001 digital network is referred to as the second generation (analog wireless was the first generation), these upgraded networks have been dubbed 2.5G (a significant notch up from the current speed and performance) and 3G. Although the timetable and technology for rolling out next-generation networks differs around the world, carriers everywhere recognize the importance of keeping today's SIM and SMS applications working during and after the upgrade. Therefore, the SIM will manage the roaming of traffic between generations of networks and between geographic locations. In addition, applications that work with today's SIM standards will be in a good position to take advantage of the higher speed and multimedia capabilities of the 3G networks as they emerge.

Carriers, mobile equipment makers, and other service providers agree that applications are the most important driver for continued growth of wireless data exchange. The providers are searching for new killer applications to generate additional revenues from their networks and increase subscriber use and loyalty. They see that individual subscribers are looking for applications that will allow them to get more from their mobile phones or wireless PDAs. Businesses need applications that make mobile employees more productive and enable them to reach their mobile customers. There are different ideas about who should develop such applications. Some carriers prefer to do their own development work, whereas others contract with third-party developers or look to the SIM and mobile equipment vendors to provide the applications. One way or another, the demand for applications continues to increase.

Wireless Application Protocol (WAP), which many people thought of as the fastest route to mobile applications, was something of a wake-up call for network operators. When wireless communications were all about voice, the operators controlled every aspect of the mobile phone. The emergence of WAP allowed well-known Web-based services like yahoo.com and literally hundreds of start-up WAP sites to download programs to the mobile handset and take control of the screen and the keypad. The wireless operators looked around and discovered that all they still really controlled was the SIM, a tiny computer deep in the guts of the mobile phone that was designed to protect security, not support applications. We'll discuss how this computer sprouted an application programming interface called the SIM Application Toolkit (SAT) and other development tools like the SIM Micro-Browser in Chapter 10, but you should know that today's SIMs are an underappreciated platform for a rich variety of mobile applications.

At the same time, application developers, especially developers who are expert in creating SMS and SIM-based applications are in short supply. It is hard to find all the information needed to start using SMS and SAT, and even harder to find clear examples of how to program specific applications. This book provides a step-by-step explanation of the commands, standards, and programming techniques that will take you from basic SMS applications to advanced SAT functionality. If you want to learn more about SMS and SIM development, this is the place to start.

Foundations and Definitions

SMS is the abbreviation for Short Message Service. SMS is a way of sending short messages to mobile telephones and receiving short messages from mobile telephones. "Short" means a maximum of 160 bytes. According to the GSM Association, "Each short message is up to 160 characters in length when Latin alphabets are used, and 70 characters in length when non-Latin alphabets such as Arabic and Chinese are used."*

The messages can consist of text characters, in which case the messages can be read and written by human beings. SMS text messages have become a staple of wireless communications in Europe and Asia/Pacific and are gradually gaining popularity in North America.

---

* GSM Association, "Introduction to SMS" on the web at http://www.gsmworld.com/technology/sms.html.

Introducing SMS and the SIM

The messages also can consist of sequences of arbitrary 8-bit bytes, in which case the message probably is created by a computer on one end and intended to be handled by a computer program on the other.

SIM is the abbreviation for Subscriber Identity Module. As its name implies, its original purpose (and continuing role) was to identify a particular mobile user to the network in a secure and consistent manner. To accomplish this, the SIM stores a private digital key that is unique to each subscriber and known only to the wireless carrier. The key is used to encrypt the traffic to and from the handset. It is essential to keep this key out of the hands of mischief makers who might get hold of a SIM and try to steal the subscriber's identity. Because smart cards were designed to be extremely difficult to crack under a variety of attacks, the smart card's core electronics and design architecture were adopted as the base of the SIM. Building applications for the SIM has a lot in common with designing smart card applications and, as we will see later, the standards that guide the evolution of smart cards and the SIM have started to converge in the international standard-setting bodies.

One of the most important standards for SIM application developers is the SIM Application Toolkit (SAT). As the name implies, the SAT standardizes the way in which applications besides the subscriber's private keys can be developed for and loaded onto the SIM. Wireless carriers are understandably sensitive about guarding the security of the SIM and preserving its primary function of subscriber identity and encryption. Because the carrier controls what code is loaded directly onto the SIM, adhering to SAT standards in building your application doesn't mean that it will run on any given network. Typically, there is a testing and certification process required for any application that is not developed directly by the network providers or SIM vendors.

On the one hand, such a process can make it difficult to get your applications on the SIM because, if any Tom, Dick, or Sally can download programs to the SIM it wouldn't be a trusted computer. On the other hand, when you do get your applications on the SIM, you will be in good company. Or, if your applications don't require the full-blown trust and security apparatus built into the SIM, you can work with SMS and a tool called the USAT Interpreter to interact with Web-based information via the SIM. As more SIMs capable of running virtual machines such as Java come to market, you can also develop applications that can be downloaded over the air—as long as the application is acceptable to the wireless carrier. This book explains the range of possibilities and illustrates the steps involved in developing those possibilities.

The SIM is the smaller of two computer chips inside a GSM mobile handset. Early SIMs typically were $1/3$ million instruction per second (MIP) with 3K memory, and most SIMs in use today are $1/2$ MIP with 16K memory. To handle virtual machines and larger applications, the current high-end SIM provides 32K of memory, with 64K SIMs anticipated within the next year. The computer chip that runs the handset is much larger, typically with a couple of megabytes of memory and a couple of MIPs of computer power. The larger chip controls the keypad and the display, encodes and decodes voice conversations, and runs the protocols that enable the handset to connect to the telephone network. The SIM may be a small computer compared with the handset computer and a tiny one compared with PDA and notebook processors, but its size doesn't have to be a gating factor for innovative applications. In fact, the SIM has about the same computing power as the first IBM PC and that computer opened the eyes of corporations and individuals to the potential of word processing, spreadsheets, and other applications to change the way we do our work and live our lives.

Bear in mind that there are other ways of exchanging data with a mobile telephone that are not covered in the following chapters. General Packet Radio Services (GPRS) is one example. There are also other ways to build mobile applications. WAP is one of the best known and has a large following. Nevertheless, SMS and the SIM have some characteristics that make them attractive for many types of application.

SMS is cheap, always on, gets through when other messages don't, is a store-and-forward system and is quite easy to build with. The SIM is portable so you can move it from one mobile device to another; it is tamper resistant, so it can be used to hold sensitive data; and it provides access to the full range of capabilities of the handset. One sweet spot for applications using SMS and the SIM is trusted transactions. Although this includes mobile commerce and financial transactions, the trust inherent in the SIM can be leveraged to a much broader group of applications where privacy and performance are important. The case-study chapters describe how companies and carriers are using this trust in real-world situations.

An SMS message nearly always gets through. If the mobile phone isn't on when you send a message, the system holds it until the phone is turned on and then delivers it. The system also can generate a return receipt that tells you that the message has been delivered. SMS messages are encrypted, so there is no fear that your message will be snatched out of the air and read. You can even add your own encryption to an SMS message so that not even the phone company can read what you are sending. There are many standards, software packages, and service providers that make building industrial-strength SMS applications easy, quick, and even fun (if you have a somewhat distorted sense of fun).

SMS and SIM in the Network Context

Before we plunge into the details of development, it is important to understand the network context in which SMS communicates with the SIM and the mobile device. The dynamic duo of SMS and SIM works as follows. The part of your application on your desktop computer or corporate server creates an SMS message to be sent to the part of your application on the mobile. This message is handed off to the short messge center of your local telephone company with the telephone number of the mobile you want it sent to. The telephone company finds the mobile and passes the SMS message to it. The message has a flag set in it that tells the handset to pass the message to the SIM. The message also has a flag that says which application on the SIM should receive the message. When the SIM receives the message from the handset, it checks to see which application to give it to and hands it off to the mobile side of your application. Figure 1-1 illustrates the flow of traffic.

Receiving a message works exactly the same way, only in reverse. The mobile side of your application generates an SMS message, attaches the telephone number of your air modem, and hands it over to the handset. The handset passes it to the network that delivers it to your desktop.

Ideally, getting an application on the air would be simply a matter of writing the two sides of your application, the server side and the mobile side, and following the appropriate standards and using a language and a runtime library of your choice. However, things are never quite that simple in the world of wireless applications.

What we'll discover is that there is a welter of options and alternative implementation possibilities available. Further, even though the mobile networks are perfectly interoperable when it comes to voice, this is far from the case when it comes to data. You certainly won't be able to move your SMS/SIM applications from one telecom operator to another as easily as moving your applications from one portal to

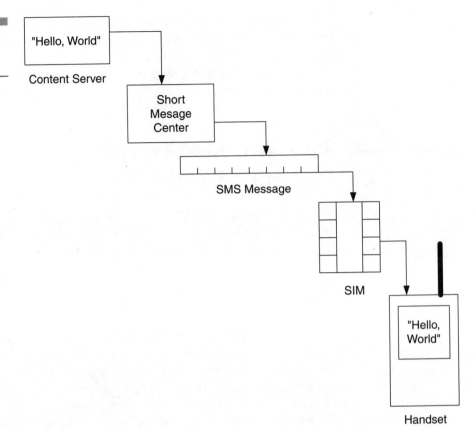

Figure 1-1
Message flow from server to screen.

another or from one Internet service provider (ISP) to another. An application that might work perfectly when both its parts are connected to the same operator might not work if the mobile part wanders off to another operator. Or, an application might work fine on one network and not at all in another.

From an application developer's perspective, such possibilities mean you have to be resourceful. You need to be able to figure out how to develop applications that fit into the wireless network according to the level of trust and security that they require and the amount of interaction and support that they need from the various points on the wireless network. There are a number of ways to proceed and choosing the right one for a particular application means being familiar with all the options.

It is important to keep in mind that the mobile network is not like the Internet technically or philosophically. The wireless operators have paid

Introducing SMS and the SIM

a great deal of money for their spectrum licenses and have invested yet more billions in transmission facilities. They care a lot about who uses their networks and for what purpose, and they often see themselves as gatekeepers in a literal sense when it comes to applications. The carriers own the spectrum and they control the SIM and, given its security requirements, they are understandably protective of it. This doesn't mean, however, that developers face an impossible hurdle getting their SIM applications on the air. Faced with the need to provide more revenue-generating services to justify the investment in next-generation networks, the carriers are eager for value-added applications and are coming to terms with the fact that internal application development is not the answer. For these reasons, carriers are increasingly open to applications that are designed to work within the SMS and SAT framework. Let's get down to the details of how to make that happen, starting with a discussion of protocol stacks and standards in the wireless network context.

Protocol Stacks

You've heard about TCP/IP and HTTP and other communication protocols, and you've probably even worked with them, but you probably haven't had to be too concerned about the details of those protocols or how they work together because there are high-level application programming interfaces to the Internet that let you ignore all the nasty details of Internet piping. This definitely isn't the state of affairs when it comes to building mobile applications.

One thing to keep in mind as you read this book is that network protocols encapsulate one another, just like those Russian dolls. Each protocol takes what it gets, puts it into an envelope with instructions written on the outside, and hands the envelope to the next guy. When the envelope gets to the other end, the receiving side of the protocol opens the envelope and passes the contents on in accordance with the instructions written on the outside of the envelope.

This process of encapsulation and de-encapsulation can be diagrammed a number of different ways. All the diagrams tell the same story. Figure 1-2 provides a simple illustration to fix the key elements of protocol encapsulation in your mind.

What makes building mobile communications different from building Internet applications is that you have to be concerned with all the envelopes, not just the first and last one.

Figure 1-2
Protocol encapsulation

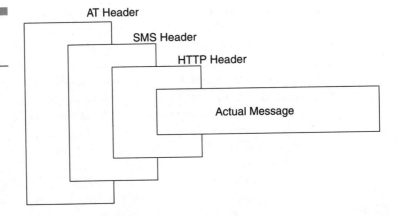

In Internet computing you only have to be concerned with the outermost envelope. You use your e-mail program to create an envelope around some text, write sguthery@mobile-mind.com on the outside, and hit SEND. Even though you may be subconsciously aware that this envelope gets put into another envelope and that into another envelope, you certainly don't worry about the details of those other envelopes. Somehow all those envelopes get your message to sguthery's mail box at mobilemind.com. At the far end I click on your envelope and my e-mail program opens it up and displays your message. I didn't worry about all the envelopes any more than you did.

What made the Internet message so easy to send is that all the nodes along the way helped out. Your computer added an envelope, your ISP added an envelope, and maybe even the network that your ISP connects to added an envelope. Everybody did his or her bit to get your message through. Those readers with long memories might remember the days when e-mail had to be routed through the Internet. E-mail addresses looked like this: sguthery!watertown!boston!rcn!uunet. All this routing is now done by the network.

The mobile network is like the early days of the Internet. The application has to be concerned with multiple envelopes. Some of these envelopes steer your SMS message through the network to the mobile device, others correctly process it on the handset, and others correctly handle it on the SIM. If you are not careful to remember how each segment follows the other, you can easily forget who you are talking to and what you are trying to say.

In some ways, the sequence and relationship of the protocols required for SMS routing are similar to the different combinations of

Introducing SMS and the SIM

numbers we have learned to dial to work our way through fixed-line voice communications. Let's say that Sally Green has just arrived at her hotel in Tokyo and wants to leave a message at her home office confirming her schedule of meetings. Sally will have to dial a string of numbers that "talk" to different parts of different phone networks. The string would look something like this:

00	to connect with the hotel switchboard
010	to reach an outside line in Tokyo
123 456 7890	to reach the local access number for Sally's international long-distance provider
54321	to verify Sally's identity with her personal identification number so the provider will put the call through
1 617	to reach the United States and Boston area code
234 5678	to reach Sally's company headquarters
200	to reach the individual to whom Sally wants to talk

This type of sequencing will be required for our mobile messages except that the numbers will be much longer, have infinitely more details, and be wholly unfamiliar to you. As we provide examples in the following chapters, we will try to keep running track of where in the hierarchy we are, whom we are talking to, and what we are trying to get them to do for us.

The Role of Standards

Communication networks by definition are governed by, paced by, and driven by standards. This makes perfect sense. If you and I don't agree completely on what bit 53 means, then when I set and hand it to you, you won't do what I thought you were going to do. There are thousands of mobile network standards. Many of them are on the Internet and free from the organizations dedicated to setting and evolving the standards, and others you have to pay a fee to obtain. Fortunately, we will be dealing with only a small percentage of the total body of mobile standards and almost all the ones we'll be talking about are free (Figure 1-3). More information about the interrelationship of the various standard-setting bodies and pointers to the sources of all of the standards mentioned in this book and many other useful standards can be found in Appendix A.

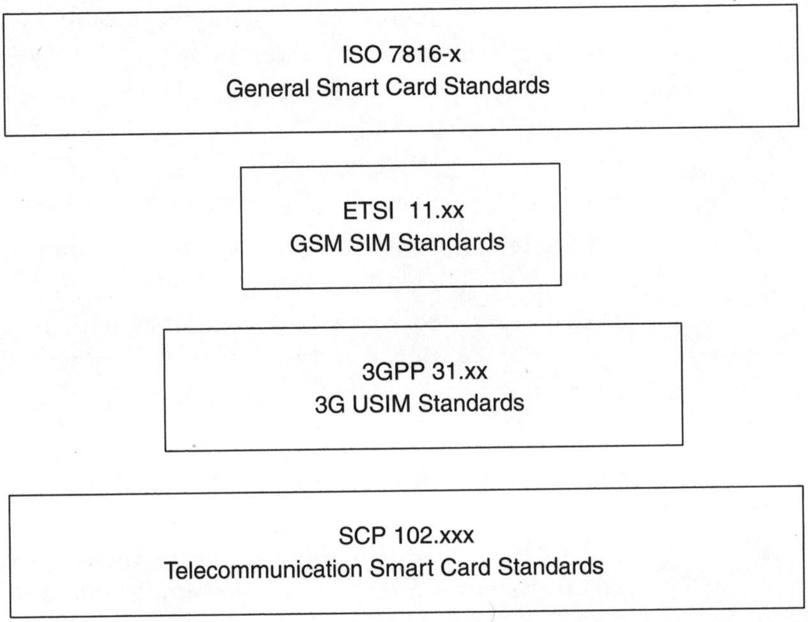

Figure 1-3
Standards suites governing the SIM.

```
ISO 7816-x
General Smart Card Standards
```

```
ETSI 11.xx
GSM SIM Standards
```

```
3GPP 31.xx
3G USIM Standards
```

```
SCP 102.xxx
Telecommunication Smart Card Standards
```

There are seven standards with regard to SMS messages that are discussed in greater depth in Chapters 2 and 3. Two govern talking to a mobile phone connected to a desktop computer:

- **3GPP 27.005**—Use of the Data Terminal Equipment—Data Circuit Terminating Equipment (DTE-DCE) interface for SMS and Cell Broadcast Service (CBS)
- **3GPP 27.007**—AT command set for 3G User Equipment (UE)

One tells us how to talk directly to the network operator:

- **3GPP 23.039**—Interface protocols for the Connection of Short Message Centers (SMSCs) to Short Message Entities (SMEs)

If we are building small applications, then we can use a mobile phone as a kind of air modem to send our messages from our computer to the phone company. If we are building a corporate application, we probably want to talk directly to the phone company over a landline.

Introducing SMS and the SIM

Two more standards give us all the nitty-gritty details about SMS messages:

- **3GPP 23.040**—Technical realization of SMS
- **3GPP 24.011**—Point-to-Point (PP) SMS support on mobile radio interface

We will spend most of our time in the first section on SMS dealing with those two in detail.

The final two provide some details about to write those messages:

- **3GPP 23.003**—Numbering, addressing, and identification
- **3GPP 23.038**—Alphabets and language-specific information

The GSM and 3GPP networks are global and must be strict about alphabets and languages. Further, the mobile network is an overlay of the landline network, which in turn grew by and large inside national boundaries. The mobile engineers couldn't just say, "Tear it down and let's start over." They had to build on what they had, so numbering schemes contain echoes from the past.

As a result, you will have to specify the properties of your messages such as which alphabet to use, how to encode that alphabet, and which numbering scheme should govern the telephone number. You'll wonder why the folks who designed the system didn't just pick one and be done with it. The reason is that the global system evolved by connecting many local systems without the benefit of a homogenizing architecture such as the Internet. The analogy is having to know the Ethernet address of the computer you want to send a packet to ... only worse.

The next set of standards govern the computer to which we are sending our SMS messages, the SIM. The SIM is essentially a smart card shorn of its plastic (i.e., just the smart part of the smart card), so it is not surprising that SIM standards are offsprings of smart-card standards.

The three basic smart card standards are:

- **ISO/IEC 7816-4**—Integrated Circuit(s) Cards (ICC) with contacts, Part 4: Interindustry commands for interchange
- **ISO/IEC 7816-8**—ICC with contacts, Part 8: Security-related interindustry commands
- **ISO/IEC 7816-9**—ICC with contacts, Part 9: Additional interindustry commands and security attributes

The two SIM standards that are derived from those and that we will be concentrating on are:

- **ETSI TS 102.221**—Smart cards; UICC—terminal interface; physical and logical characteristics
- **ETSI TS 102.222**—ICC; administrative commands for telecommunications applications

These standards describe the SIM platform. Think of them as the documentation for an Intel processor with the Win32 application programming interface (API). On top of this platform, we will consider two programming metaphors: a microbrowser metaphor and an executable program metaphor.

The SIM microbrowser is something of a misnomer. It isn't an attempt to turn the SIM into a Web browser. Rather, it is a byte-code interpreter that allows the SIM to download, display, interact with the subscriber, and communicate with your application with a Web-based set of instructions and then throw away the instructions once the interaction has been completed. The byte-coded instructions that are sent to the microbrowser are very much like the pages that are sent to your desktop browser, which is the reason for the name "microbrowser," however inaccurate technically.

This "fire-and-forget" model of user interaction fits very well with the constraints and capabilities of the SIM. As a result, many network operators favor the SIM microbrowser as a more lightweight and easily controlled way to get value-added applications to their customers than the more ponderous and administratively expensive executable program model.

There is one key standard that describes the SIM microbrowser, now called the USAT Interpreter:

- **3GPP TS 31.113**—USAT interpreter byte codes

The second, less widely used, model of computation is where you install your application code directly on the SIM just as you might install a new program on your PC. Your own experience has probably taught you that you are more likely to run into trouble installing a program on your computer (sometimes call *applets*) than simply viewing a page on the Web, and this has been the experience of the network operators, except that they deal with millions and tens of millions of computers using their customers' SIMs, so having trouble installing a program is serious problem for them.

Building an executable program for the SIM is much more complex than simply sending pages to a program already installed on the SIM, as with the microbrowser. As a result, there are more standards that govern this type of application development. The ones that we will consider are:

- **ETSI TS 02.19**—Subscriber Identity Module Application Programming Interface (SIM API): Service description, stage 1
- **ETSI TS 03.48**—Security mechanisms for SAT, stage 2

Whether you are building a microbrowser application or an executable program application, your code is written against an API inside that SIM. This interface is described in the last standard we will be using:

- **ETSI TS 102.223**—Smart cards; card application toolkit

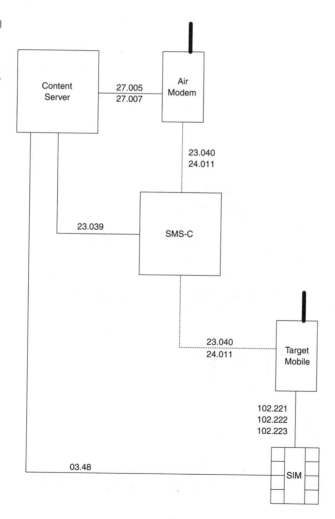

Figure 1-4
Standards on mobile application interfaces.

Preview of Coming Chapters

Not surprisingly given the title, this book is divided into two major sections: SMS messaging and SIM application programming. The SIM section is divided further into two parts, one on the SIM microbrowser and the other on SIM applets.

In the SMS section we focus on getting an SMS message to the mobile and handling an SMS message that is sent from the mobile. Because our primary concern is working with the 3G system to get the message there and get it back, we won't worry too much about what the message contains and will use simple text messages in most of our examples.

In the microbrowser part of the section on the SIM microbrowser, we use techniques discussed in the SMS section to send Internet-style Web pages to the SIM. These pages are rendered by the microbrowser SIM and then deleted. There is a surprising range of mobile applications you can build with this seemingly modest capability. We explore some of those possibilities in detail.

In the SIM applets section, we discuss installing permanent applications on the SIM. This can be done when the SIM is manufactured or can be done later after the SIM is in use in your mobile phone. Because the amount of memory on the SIM is quite limited, you have to work closely with the network operator to use SIM applets.

There are three case study chapters, Chapters 6, 11, and 13. The case studies are of increasing complexity and show how wireless carriers, corporate customers, and third-party application developers use the techniques in the preceding chapters to bring successful applications into being. The cases also illustrate how SMS and SIM applications add value for the operations and wireless customers.

Summary

If you'll pardon the pun, there are lots of moving parts in building mobile applications. Building a killer mobile application isn't a matter of slinging millions or even thousands of lines of code. It's scalpel, not machete, work.

You're probably used to working with large objects like Windows COM objects or SQL databases or maybe CICS transactions. Rarely if ever do you contemplate the bits and bytes that form the base on which you're building. That is not the case with mobile applications. There is such incredible pressure on all the technical dimensions of mobile computing, for example, bandwidth, battery life, weight, cost, and transmission time, that no effort has been spared to squeeze the last little bit of value out of every little bit. What might be casually allocated to an 8-byte or a even a 32-bit word on a desktop computer, a TRUE/FALSE flag, for example, will be given exactly 1 bit in mobile computing and then only after it convinces everybody that it really needs to exist.

It may seem strange and even frustrating at first, like painting with a one-bristle paintbrush. Succeeding in this space-constrained and absolutely precise world of mobile programming requires a different set of skills and tradeoffs than building applications for the desktop or even the PDA. After a while, you will discover that, once you learn the colors and the techniques, you can make very impressive—and functional—pictures. So let's begin building assembly language programs for the biggest computer in the world, the worldwide telecommunications network.

CHAPTER 2

Basic SMS Messaging

There are many software development kits and products on the market that you can use to connect your application to SMS messaging. These range is from very low-level packages that simply connect a serial line port to the mobile phone up to all-singing, all-dancing packages that provide all sorts of message management services. In between are packages that provide various APIs to SMS messaging such as Telephone Aplication Program Interface (TAPI) that make it easy to integrate SMS messaging into existing application suites.

We will begin with basic, low-level messaging and work our way up the food chain. You may never actually build an application using these low-level commands but it's good to know what's under the hood and what's possible just in case you get stuck and have to reach for the spanners. The higher-level packages are essentially fancy ways of generating those low-level commands.

In the next couple of paragraphs we discuss setting up your mobile application development workbench.

Connecting the Handset

Every GSM and 3GPP handset is an air interface modem and a plain old telephone handset. This means you can connect the handset to an external interface on your computer and send it AT commands just as you did with your dial-up modem. The physical connection can be any one that your computer offers such as a serial port, a USB port, or an IrDA port. We are going to use a serial COM port for the examples in this chapter because it is the most widely used one at present.

Besides an activated GSM phone you'll need a cable that connects the phone and the serial port on your computer. You'll also need to install a modem driver on your computer that knows how to talk to the phone. The cable and the driver depend on the model of the handset you are using. Most handset manufacturers offer a data kit of some sort for their handsets that includes the right cable and the driver. Examples in this chapter use a Nokia 5190 handset and the SoftRadius driver and cable for that handset from Option Inc. Nokia produces several very nice data kits called the Nokia Data Suite and the Nokia PC Connectivity SDK, which accomplish the same thing.

After you've installed the driver, you can use the same terminal program that you use for dial-up modems to test the connection. On a Windows system, just use HyperTerminal. Type "AT" on the COM

Basic SMS Messaging

port connected to the phone. If everything is working properly, you should see "OK." Now you're ready to start building SMS applications.

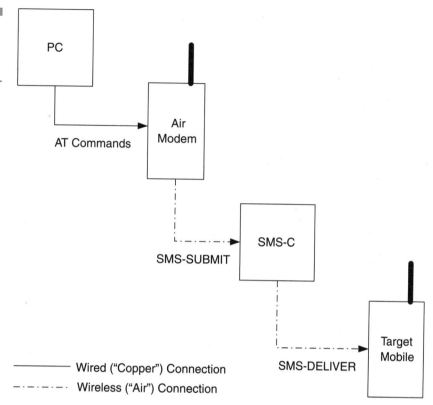

Figure 2-1
Message flow from desktop PC to mobile handset.

Communicating with the Handset

In addition to many of the standard V.32ter and Hayes modem dial-up modem AT commands, your mobile handset supports a set of AT commands that are particular to connecting to the GSM network and sending short messages. If you're a gnarly old Hayes modem hacker, you'll feel right at home. The standard handset AT commands are described in the following two documents:

- **3GPP 27.005**—DTE-DCE interface for SMS and CBS
- **3GPP 27.007**—AT command set for 3G UE

The big difference between using a dial-up modem connected to the landline telephone network and a handset connected to the GSM network is how much you can see and say to the network itself. About the only thing you said to the network through your dial-up modem was "Connect me to the following number." You did this with the Hayes ATDT command.

```
ATDT 6172345678
```

This caused the dial-up modem to generate the right dual tone mulit-frequency (DTMF) tones on the line to cause the telephone network to set up a dedicated circuit connection between your modem and the modem that answered at the other end. Once the connection was established, all the wired network did was move an analog signal from one end to the other. The modems on both ends took care of turning the analog signal into bits, frames, packets, and messages.

A mobile network is continuously and more intimately involved in the bit stream if for no other reason than the modem you are trying to communicate with—the mobile handset out there somewhere—keeps moving around.

In the 27.007 AT command set you will find some old friends such as ...

ATD	Dial command
ATE	Command echo
ATH	Hang up call
ATA	Answer call
ATS	Select an S-register
ATQ	Result code surpression
ATZ	Recall stored profile

But you'll also find lots of commands that are more about you talking to and about the network than to and about the handset modem such as ...

AT+CSCS	Select character set
AT+WS46	Select wireless network
AT+CBST	Select bearer service type

Basic SMS Messaging

AT+CRLP	Radio link protocol
AT+CR	Service reporting protocol
AT+CRC	Cellular result codes
AT+COPS	Operator selection
AT+CSCA	Service center address

Finally, because a mobile handset is a much more capable device than the old V.32 Hayes modem, there are many commands that you can use to manipulate it such as ...

AT+CPBF	Find phone book entries
AT+CPBR	Read phone book entry
AT+CPBW	Write phone book entry
AT+CMGL	List messages
AT+CMGR	Read messages
AT+CMGS	Send message

For example, after I connected my mobile phone to my PC and fired up HyperTerminal, I used AT + CMGL to get a list of the messages that were stored in the SIM:

```
AT
OK
AT+CMGL
+CMGL: 1,1,24
07919171095710F0040B917118530400F90000103080406553580 5C8
   329BFD06
+CMGL: 2,1,30
07919171095710F0040B917118530400F900001030111041805 80CC8
   329BFD6681EE6F399B0C
+CMGL: 3,1,23
07919171095710F0040B917118530400F9000010301110612558 04E5
   B2BC0C
+CMGL: 4,1,25
07919171095710F0040B917118530400F90000103011100203580 665
   79595E9603
+CMGL: 5,1,24
```

```
07919171095710F0040B917118530400F9000010301110645458 05C8
   329BFD06
+CMGL: 6,1,28
07912160130300F4040B917118530400F90000108050709244690AD4
   F29C0E8A8164A019
+CMGL: 7,1,37
07912160130300F4040B917118530400F900001080507003516914D7
   329BCD02A1CB6CF61B947FD7
E5F332DB0C

OK
```

There were seven messages stored in the SIM. In Chapter 3, we will analyze the numbers and find not only the message but also lots of interesting information about the message such as who sent it and when it arrived.

Communicating with the Network

Because the mobile network is an active participant in moving messages between your application and a mobile device, you have to be much more concerned with the details of formatting the messages you send. Remember the mobile network actually looks at the bytes in your message (actually in the headers on your message) to figure out what to do with it. "Please tell Sally Green wherever she is that dinner won't be ready until 7" just doesn't hack it.

We will discover that there are lots of things besides who should receive the message that you can tell the GSM network and its SMS centers (SMSC). The string of bytes that you send into the network contains not only the message but also lots of other information that instructs the network as to how and when you want this to happen.

The two standards that govern the construction of SMSs what we will be using are:

- **3GPP 23.040**—Technical realization of SMS
- **3GPP 24.011**—PP SMS support on the mobile radio interface

These standards cover the encoding of the message that gets delivered to the destination handset and the encoding of the instructions to the GSM network and the SMSC.

Basic SMS Messaging

Remember our discussion in Chapter 1 about the encapsulation of protocols? In building low-level commands for sending SMSs, we are in fact talking to three separate entities: the local handset to which we are sending AT commands, the network and its SMSC, and the endpoint mobile that will receive the message.

Figure 2-2 shows the complete SMS header diagram. We are covering only the outermost two in this chapter and will get to the others in later chapters. You build all the headers, so you will have to remember whom you are talking to and what you are saying to them as you build your SMS message.

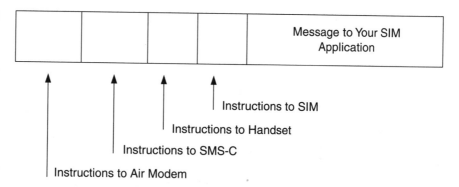

Figure 2-2
SMS message headers.

Hello, Mobile World

Let's start by opening a serial port connection to the local handset. My Nokia 5190 is connected to COM5, so using the C programming language I'd write:

```c
handle = CreateFile("COM5",
     GENERIC_READ | GENERIC_WRITE, // read and write
     0, // exclusive access
     NULL, // no security
     OPEN_EXISTING,
     0, // no overlapped I/O
     NULL); // null template
```

It is on this connection that we will send AT commands to the local handset that in turn will relay the information to the GSM network.

We must set this serial connection to binary so that the operating system and its drivers don't touch the data as it passes through, for example, by adding carriage returns and line feeds. We want the data we construct to get to the handset and to the network exactly as we built it and not with any "help" from folks along the way.

How this is done changes from handset driver to handset driver. For the particular driver I'm using, binary information sent on this connection is hex-encoded as ASCII characters, so if you wanted to send the byte 0x9D, you'd send the ASCII string "39 44": 39 is the hexadecimal value for the ASCII character 9 and 44 is the hexadecimal value for the ASCII character D.

Let's start by sending a simple "Hello, world" message to the mobile phone at +1 617-230-1346.

What we do is pack in a hex-encoded byte blob all the information needed to get this message to its destination along with the message itself and ship this blob off to the carrier's SMSC which in turn will get it to where it is going.

The byte blob is an SMS_SUBMIT Transfer Protocol Data Unit (TPDU). We'll take a detailed look at TPDUs in Chapter 3. The one at hand consists of the following fields:

1. Transfer protocol parameters = 0x01 (an SMS_SUBMIT TPDU)
2. Message reference number = 0x00 (let the handset assign it)
3. Length of destination number in digits = 0x0B (11 digits)
4. Type of destination number = 0x91 (international format)
5. Destination telephone number (nibble swapped) = 0x6171321043F6
6. Protocol identifier = 0x00 (implicit)
7. Data coding scheme = 0x00 (GSM default alphabet)
8. Message length = 0x0C (there are 12 characters in "Hello, world")
9. Message = 0xC8329BFD6681EE6F399B0C ("Hello, world")

The coding of the actual message, "Hello, world," requires some explanation. No stone is left unturned when it comes to optimizing the use of the air interface. If we had transmitted the ASCII characters as bytes, we would have wasted a bit for every character we sent because ASCII characters are coded on 7 bits and sending this message as an 8-bit byte wastes 1 bit. Now, 1 bit is no big deal if you have megabytes of memory and gigabytes of disk space, but on an air interface this represents a waste of one-eighth of the channel capacity and this cannot be tolerated.

Basic SMS Messaging

What we do is very simple. First, we put the first character into the first byte. Next, we take the low-order bit of the seven bits of the second ASCII character and stuff it in the unused high-order bit of the first byte. Now we put the six remaining bits of the second character into the second byte. Next, we take the two low-order bits of the seven bits of the third ASCII character and stick them into the two unused high-order bits in the second byte, and so forth.

Here's the result of applying this packing algorithm to "Hello, world":

	H	e	l	l	o	,		w	o	r	l	d
Unpacked	48	65	6C	6C	6F	2C	20	77	6F	72	6C	64
Packed	C8	32	9B	FD	66	81	EE	6F	39	9B	0C	

In this case, the low-order bit of the ASCII character "e" is 1, so we put that into the high-order bit of the first byte, which is unused after we put the seven bits of "H" into the low-order bits of the byte. This turns 0x48 into 0xC8. Now, we take the remaining low-order six bits of "e," 0x25, and put them into the low-order six bits of the second byte. We then put the two low-order bits of the seven bits of the ASCII character "l," namely 0 and 0, into the two unused high-order bits of the second byte. This yields a second byte of 0x32, and so forth. As a result, we save a complete byte.

Here is some C code that performs this packing and also makes the hex-encoded ASCII characters that are sent to the handset:

```
#include <string.h>

void unpack78(char *p, int n, char *s);
void pack78(char *s, char *p, int *n);

#define N(c)  (c<=0x39?((c)-0x30):((c)-0x37))
#define M(c)  (c<=0x09?((c)+0x30):((c)+0x37))

void pack78(char *s, char *p, int *n)
{
  unsigned char byte[160];
  int bits, i, j, k;

  k = strlen(s);
```

```c
/* Pack the ASCII characters into bytes */
for(i = j = bits = 0; i < k; i++, j++) {
  if(bits == 7) {
    bits = 0;
    i++;
  }
  byte[j] = (s[i]&0x7F)>>bits | s[i+1]<<(7-bits);
  bits = (++bits)%8;
}

/* Convert bytes to ASCII nibbles */
for(i = 0; i < j; i++) {
  k = (byte[i]>>4)&0x0F;
  *p++ = M(k);
  k = byte[i]&0x0F;
  *p++ = M(k);
}

  *n = j;
}
```

For the sake of completeness, here is the corresponding unpacking routine that we will need when we receive messages from the handset:

```c
void unpack78(char *p, int n, char *s)
{
  int bits, i, j;
  unsigned c, byte[160];

  /* Convert ASCII nibbles to bytes */
  for(i = 0; i < n; i++) {
    byte[i] = N(*p);p++;
    byte[i] = (byte[i]<<4)|(N(*p));p++;
  }

  /* Extract the ASCII characters from the bytes */
  for(i = j = bits = 0; j < n; i++, j++) {
    if(bits > 0)
      c = byte[i-1]>>(8-bits);
    else
```

Basic SMS Messaging

```
            c = 0;
         if(bits < 7)
            c |= byte[i]<<bits;
         *s++ = c&0x7F;
         bits = (++bits)%8;
         if(bits == 0)
            i -= 1;
      }
      *s = '\0';
   }
```

So the complete SMS_SUBMIT TPDU for "Hello, world" looks like this:

01000B916171321043F600000CC8329BFD6681EE6F399B0C

All we have to do now is use an AT command to send this TPDU off to the SMSC. This is the send-message AT command:

AT+CMGS=<TPDU length><CR><SMSC address><TPDU><CTRL-Z>

The SMSC address is the telephone number of the SMSC to which the handset should send the TPDU. Like the destination telephone number, the telephone number of the SMSC consists of three sub-fields:

1. Length of the telephone number in octets = 0x07 (7 octets)
2. Format of the SMS telephone number = 0x91 (international format)
3. Telephone number of SMSC (nibble swapped) = 0x9171095710F0

This particular SMSC is in the VoiceStream network, where the handset set I am using as my air modem is registered. When you test this example, you'll have to replace this phone number with the phone number of the SMSC in the network to which you subscribe.

The following code is what I write to COM5 to send "Hello, world" to my mobile:

```
AT+CMGS=24
07919171095710F001000B916171321043F600000CC8329BFD6681EE
   6F399B0C^Z
```

The <CR> is the byte 0x0D and the CTRL-Z is the byte 0x1A in the actual sequence of bytes sent to the handset. Give it a try. I look forward to receiving your SMS!

So what happens if the mobile we send a message to sends one back? We can list all the messages in the handset using the AT command:

AT+CMGL

and then we can retrieve the one we want by using the AT command:

AT+CMGR=<index>

When using AT+CMGR to retrieve the latest message that arrived, the handset replies with:

+CMGR: 1,21
07919171095710F0 040B916171321043F6 00 00 10201180234458
  02C834

The 1 on the first line indicates the status of the message. The number 1 means this is a received message that has been read. The following 21 is the number of bytes in the TPDU in the following data. The following line shows the data comprising the message. It is in the same general format as the data in the AT+CMGS command, namely the SMSC telephone number followed by a TPDU. In this case, however, it is the telephone number of the SMSC delivering the message and an SMS_DELIVER TPDU rather than an SMS_SUBMIT TPDU. In other words, the TPDU is being delivered to the handset rather than the handset submitting a TPDU to the network. We'll discover that what is delivered is not exactly the same as what is submitted.

The SMSC phone number is just like the one we sent to, so let's analyze the SMS_DELIVER TPDU. We'll be using 3GPP TS 23.040 to do this.

1. Transfer protocol parameters = 0x04 (SMS_DELIVER with no more coming)
2. Length of originating address = 0x0B (11 bytes)
3. Type of originating address = 0x91 (international format)
4. Originating address (nibble swapped) = 0x6171321043F6 (+1 617 230 1346)

5. Protocol identifier = 0x00
6. Data coding scheme = 0x00
7. Service center timestamp (nibble swapped) = 0x10 20 11 80 23 44 58 (Y/M/D/H/M/S/Zone = 2001 February 11, 8:32:44, GMT-5)
8. Length of message = 0x02 (2 bytes)
9. Message = 0xC834 ("Hi")

To unpack 0xC834, just take the top bit from 0xC8 and put it to the right of 0x34. This turns 0xC8 into 48 and 0x34 into 0x69.

	H	i
Packed	C8	34
Unpacked	48	69

Therefore, at 8:32 Eastern Standard Time the mobile phone +1 617 230 1346 sent me the message "Hi." I'm a fascinating conversationalist when I'm talking to myself.

You can achieve a variety of special effects by setting various parameters in the SMS_SUBMIT TPDU. We will cover TPDUs in detail in Chapter 3, but to give you a feel for what is possible, suppose we'd like to have our message pop up on the screen and at the same time provide a way to quickly open up a telephone call back to the sender. This is useful for sending alerts needing an immediate response. All the recipient has to do is pick "Use Number" or "Return Call" or whatever phrase his or her handset uses to indicate the presence of a return call number in the SMS.

The pop up message is accomplished by setting the Data Coding Scheme byte to 0xF0 rather than to 0x00. This says "put the message on the screen don't just store it in the SIM" and tells the user that a new message has arrived. Creating a return call path is accomplished by setting the Protocol Identifier to 0x5F rather than to 0x00. Give it a try.

You can see why we call SMS messaging the assembly language programming of the wireless network. An SMS header is in a very real sense an instruction for a very large instruction word (VLIW) computer where the computer is the mobile telephone network. Every bit counts and these bits interact. Further, the network executes many of the fields of the instruction in parallel as if it were horizontal microcode. If you missed the era of bit-slice computers, now's your chance.

There are many other AT commands that you can send to the handset. We can't go into all of them here. You can download 3GPP TS 27.005 and 3GPP TS 27.007 for the complete story. All phones that support data support the major sending and receiving AT commands. You'll have to experiment to find out which of the more esoteric commands such as AT+CUSD (unstructured supplementary service data) and AT+CMER (mobile equipment event reporting) are supported with your driver and phone combination.

Table 2-1 lists some of the error codes you might run into when an AT command returns an error.

TABLE 2-1

SMS Error Codes

Error	Error Meaning
0-127	GSM 04.11 Annex E-2 values
128-255	GSM 03.40 section 9.2.3.22 values
300	Phone failure
301	SMS service not available
302	Operation not allowed
303	Operation not supported
304	Invalid PDU mode parameter
305	Invalid text mode parameter
310	SIM not inserted
311	SIM PIN needed
312	SIM PIN2 needed
313	SIM failure
314	SIM busy
315	SIM incorrect
320	Memory failure
322	Memory full
331	No network
332	Network timeout
500	Unknown error
512	Manufacturer specific error

Basic SMS Messaging

Table 2-2 gives the GSM 7-bit default alphabet as specified by 3GPP 23.038 and the corresponding ISO-8859 decimal codes where applicable.

TABLE 2-2 SMS 7-Bit Default Character Encoding

Hexameter	Decimal	Character name	Character	ISO-8859 Decimal
0x00	0	Commercial AT sign	@	64
0x01	1	British monetary unit—pound	£	163
0x02	2	US monetary unit—dollar	$	36
0x03	3	Japanese monetary unit—yen	¥	165
0x04	4	Lowercase e with accent grave	è	232
0x05	5	Lowercase e with accent acute	é	233
0x06	6	Lowercase u with accent acute	ú	250
0x07	7	Lowercase i with accent grave	ì	236
0x08	8	Lowercase o with accent grave	ò	242
0x09	9	Uppercase C with cedilla	Ç	199
0x0A	10	Line feed (\n)		10
0x0B	11	Uppercase O with stroke	Ø	216
0x0C	12	Lowercase o with stroke	ø	248
0x0D	13	Carriage return (\r)		13
0x0E	14	Uppercase A with ring	Å	197
0x0F	15	Lowercase a with ring	å	229
0x10	16	Uppercase Greek delta	Δ	
0x11	17	Underline character	—	95
0x12	18	Uppercase Greek phi	Φ	
0x13	19	Uppercase Greek gamma	Γ	
0x14	20	Uppercase Greek lambda	Λ	
0x15	21	Uppercase Greek omega	Ω	
0x16	22	Uppercase Greek pi	Π	
0x17	23	Uppercase Greek psi	Ψ	
0x18	24	Uppercase Greek sigma	Σ	

continued on next page

TABLE 2-2 SMS 7-Bit Default Character Encoding (continued)

Hexameter	Decimal	Character name	Character	ISO-8859 Decimal
0x19	25	Uppercase Greek theta	Θ	
0x1A	26	Uppercase Greek xi	Ξ	
0x1B	27	Escape to extension table		
0x1B0A	27 10	Form feed (\f)		12
0x1B14	27 20	Circumflex	^	94
0x1B28	27 40	Left curly bracket	{	123
0x1B29	27 41	Right curly bracket	}	125
0x1B2F	27 47	Backslash	\	92
0x1B3C	27 60	Left square bracket	[	91
0x1B3D	27 61	Tilde	~	126
0x1B3E	27 62	Right square bracket	]	93
0x1B40	27 64	Vertical stroke	|	124
0x1B65	27 101	Euro sign		164
0x1C	28	Uppercase AE	Æ	198
0x1D	29	Lowercase ae	æ	230
0x1E	30	Lowercase German ss	ß	223
0x1F	31	Uppercase E with circumflex	Ê	202
0x20	32	Space		32
0x21	33	Exclamation mark	!	33
0x22	34	Question mark	?	34
0x23	35	Hash sign	#	35
0x24	36	General currency sign	¤	164
0x25	37	Percent sign	%	37
0x26	38	Ampersand	&	38
0x27	39	Apostrophe	'	39
0x28	40	Left parenthesis	(	40
0x29	41	Right parenthesis	)	41
0x2A	42	Asterisk	*	42

continued on next page

Basic SMS Messaging

TABLE 2-2 SMS 7-Bit Default Character Encoding (continued)

Hexameter	Decimal	Character name	Character	ISO-8859 Decimal
0x2B	43	Plus sign	+	43
0x2C	44	Comma	,	44
0x2D	45	Hyphen and minus sign	-	45
0x2E	46	Period	.	46
0x2F	47	Slash	/	47
0x30	48	Digit 0	0	48
0x31	49	Digit 1	1	49
0x32	50	Digit 2	2	50
0x33	51	Digit 3	3	51
0x34	52	Digit 4	4	52
0x35	53	Digit 5	5	53
0x36	54	Digit 6	6	54
0x37	55	Digit 7	7	55
0x38	56	Digit 8	8	56
0x39	57	Digit 9	9	57
0x3A	58	Colon	:	58
0x3B	59	Semicolon	;	59
0x3C	60	Less-than sign	<	60
0x3D	61	Equal sign	=	61
0x3E	62	Greater-than sign	>	62
0x3F	63	Question mark	?	63
0x40	64	Inverted exclamation mark	¡	161
0x41	65	Uppercase A	A	65
0x42	66	Uppercase B	B	66
0x43	67	Uppercase C	C	67
0x44	68	Uppercase D	D	68
0x45	69	Uppercase E	E	69
0x46	70	Uppercase F	F	70

continued on next page

TABLE 2-2 SMS 7-Bit Default Character Encoding (continued)

Hexameter	Decimal	Character name	Character	ISO-8859 Decimal
0x47	71	Uppercase G	G	71
0x48	72	Uppercase H	H	72
0x49	73	Uppercase I	I	73
0x4A	74	Uppercase J	J	74
0x4B	75	Uppercase K	K	75
0x4C	76	Uppercase L	L	76
0x4D	77	Uppercase M	M	77
0x4E	78	Uppercase N	N	78
0x4F	79	Uppercase O	O	79
0x50	80	Uppercase P	P	80
0x51	81	Uppercase Q	Q	81
0x52	82	Uppercase R	R	82
0x53	83	Uppercase S	S	83
0x54	84	Uppercase T	T	84
0x55	85	Uppercase U	U	85
0x56	86	Uppercase V	V	86
0x57	87	Uppercase W	W	87
0x58	88	Uppercase X	X	88
0x59	89	Uppercase Y	Y	89
0x5A	90	Uppercase A	Z	90
0x5B	91	Uppercase A with dieresis	Ä	196
0x5C	92	Uppercase O with dieresis	Ö	214
0x5D	93	Uppercase N with tilde	Ñ	209
0x5E	94	Uppercase U with dieresis	Ü	220
0x5F	95	Section sign	§	167
0x60	96	Inverted question mark	¿	191
0x61	97	Lowercase a	a	97
0x62	98	Lowercase b	b	98

continued on next page

TABLE 2-2 SMS 7-Bit Default Character Encoding (continued)

Hexameter	Decimal	Character name	Character	ISO-8859 Decimal
0x63	99	Lowercase c	c	99
0x64	100	Lowercase d	d	100
0x65	101	Lowercase e	e	101
0x66	102	Lowercase f	f	102
0x67	103	Lowercase g	g	103
0x68	104	Lowercase h	h	104
0x69	105	Lowercase i	i	105
0x6A	106	Lowercase j	j	106
0x6B	107	Lowercase k	k	107
0x6C	108	Lowercase l	l	108
0x6D	109	Lowercase m	m	109
0x6E	110	Lowercase n	n	110
0x6F	111	Lowercase o	o	111
0x70	112	Lowercase p	p	112
0x71	113	Lowercase q	q	113
0x72	114	Lowercase r	r	114
0x73	115	Lowercase s	s	115
0x74	116	Lowercase t	t	116
0x75	117	Lowercase u	u	117
0x76	118	Lowercase v	v	118
0x77	119	Lowercase w	w	119
0x78	120	Lowercase x	x	120
0x79	121	Lowercase y	y	121
0x7A	122	Lowercase z	z	122
0x7B	123	Lowercase a with dieresis	ä	228
0x7C	124	Lowercase o with dieresis	ö	246
0x7D	125	Lowercase n with tilde	ñ	241
0x7E	126	Lowercase u with dieresis	ü	252
0x7F	127	Lowercase a with grave	à	224

Summary

In this chapter we covered the low-level programming of SMS messages. We sent and received messages using AT commands to communicate with a GSM handset. We built a simple outgoing SMS message and we interpreted a simple incoming SMS message at the bit and byte levels using 3GPP standards. In the next chapter, we move beyond simple messages and explore the range of possibilities that are available with SMS message encoding.

CHAPTER 3

Details of SMS-SUBMIT and SMS-DELIVER

There are six kinds of messages that flow in the SMS network but we are going to focus on only two of them: the message that leaves your application and the message that this message causes to be delivered to the handset. What may be a bit surprising is that these messages are different. Remember that in the last chapter we found that the network is an active part of message handling and not just a passive shuffler of bits.

Of the six message types, two go from the mobile device (whether it is serving as an air modem or out walking around somewhere) into the network:

- SMS-SUBMIT "submits" a message to the SMSC, generally for forward transmission to another mobile device
- SMS-COMMAND goes to the SMSC and tells it to do something

SMS-SUBMIT is one of the two messages we will be describing in detail below.

Four of the six messages go from the network to the mobile device. Of these, only one carries a message from another mobile device:

- SMS-DELIVER delivers a message from another mobile

This is the other message we will be considering in detail. The other three messages are generated by the network itself and tell the mobile what is going on (Figure 3-1):

- SMS-SUBMIT-REPORT reports on the results of an SMS-SUBMIT or an SMS-COMMAND
- SMS-DELIVER-REPORT reports on the success or failure of the delivery of an SMS-DELIVER or SMS-STATUS-REPORT message
- SMS-STATUS-REPORT reports on the results of an SMS-COMMAND message

We are going to concentrate on SMS-SUBMIT and SMS-DELIVER because that's where the tire meets the road. These are the workhorses of SMS messaging. In general, you send messages and you receive messages. The other four commands are particularly useful in high volume situations or where you want to exercise some fine-tune control over the messaging. By the time you are in such a situation, you are probably using some high-level tools that hide the details of those messages.

We were introduced to SMS-SUBMIT and SMS-DELIVER in Chapter 2, where we sent and received the "Hello, world" message. They

Details of SMS-SUBMIT and SMS-DELIVER

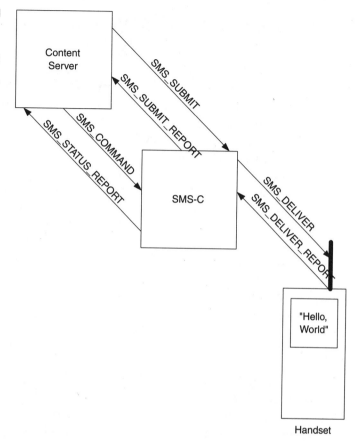

Figure 3-1
The six message types in SMS message traffic.

were the packets that contained the message itself and the number of the mobile telephone we wanted to send it to (SMS-SUBMIT) or the mobile that sent it to us (SMS-DELIVER). At the end of Chapter 2 we saw that, by changing just a couple of bytes in the SMS-SUBMIT packet, we were able to tease out some surprisingly useful behavior. We popped up our message on the screen and included a phone number that the recipient could use to initiate a call back with just a keystroke or two.

In this chapter we are going to explore all the features and wonders of the message pitcher, SMS-SUBMIT, and its catcher, SMS-DELIVER. We will concentrate on features that you can use to build interesting applications and pay less attention to features that are really there for running telephone networks. If you are building a telephone network, then you probably should be reading the standards themselves and not this exegesis.

Numbering Plans and Mobile Telephone Numbers

Let's begin with addresses. As you might expect, 3GPP standards embrace all the number plans you'd expect to find in the telephone network and probably a couple you didn't even know existed. There are two whole standards devoted solely to address formats: 3GPP TS 24.011 and 3GPP TS 29.002.

Because our focus is on what can be done with messaging and not with all the various ways of getting a message to where its going, we always are going to use the international numbering format that we saw in Chapter 2. This is the typical format and almost always works. For US numbers the international format for telephone numbers is 8 bytes long and consists of the number of digits in the phone number, 0x0B, followed by 0x91, indicating an international format number, followed by the number itself, nibble swapped. Thus +1 617 230 1346 becomes

```
0B 91 61 71 32 10 43 F6
```

Note that F is used as the filler nibble at the end of the number. (A *nibble* is 4 bits or half a byte. A filler is a bit pattern used to fill unused space, like digital Styrofoam.) In the following discussions we use an address length of 8 bytes, but the length would depend on whether the telephone numbers of other countries or other numbering plans are used.

As with most protocol packets, SMS-SUBMIT and SMS-DELIVER begin with a mandatory header that is found on all messages and then comes an optional part that is only present if the options that it describes are being used. The idea is to keep messages as short as possible but to allow all kinds of variations and features in special and usually atypical situations.

SMS_SUBMIT

Let's first take a look at the mandatory part of SMS_SUBMIT (Figure 3-2).

Details of SMS-SUBMIT and SMS-DELIVER

Figure 3-2
SMS_SUBMIT headers.

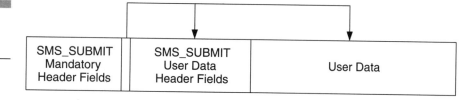

The minimal SMS-SUBMIT header consists of 13 bytes including the 8-byte destination address. The five non-address bytes are:

Field Name	Bytes	Description
SMS Flags	1	General information about the packet
Message Reference	1	Number ID of the message; typically set by handset
Protocol Identifier	1	How the SMS packet is going to be used
Data Coding Scheme	1	How the message is encoded
User Data Length	1	Length of following User Data field

The 1-byte SMS Flags field for SMS-SUBMIT consists of six subfields:

Subfield Name	Bits	Description
Message Type Indicator	2	0x01 for SMS-SUBMIT
Reject Duplicates	1	If set to 1 the SMSC should reject this message, if it is currently holding an exact copy of it
Validity Period Format	2	The optional data holds a field that tells how long to try to keep delivering this message; the field is in the format specified by these bits

continued on next page

Subfield Name	Bits	Description
Reply Path	1	If set to 1, receiving mobile should allow easy selection of sending phone number to return a message or initiate a voice call
User Data Header Indicator	1	There are optional feature descriptions after the mandatory header and before the message itself
Status Report Request	1	Set to 1 if a delivery report should be returned

The most commonly used fields from the point of view of application development are:

- **Protocol Identifier**—Just like the protocol field in IP; says what protocol is encoded in the payload
- **Data Coding Scheme**—A combination of normal data encoding and the TCP/UDP port, which determines which application handles the message
- **Reply Path**—Setting this bit activates the display and use of the originating telephone number on the receiving end
- **User Data Header Indicator**—When set to 1, this signals that there are some special features later in the packet

The part of the message after the mandatory header is called the User Data field. If the User Data Header bit is set, then the User Data field contains one or more optional feature descriptors before the message itself. The optional feature descriptors are officially called Information Data Elements. The User Data Length field in the mandatory header counts all bytes in the User Data field, i.e., the bytes in the optional feature descriptions and the bytes in the message itself.

The User Data Length is the last byte in the mandatory header. If it is not zero then the next byte is the first byte in the User Data Field. If there are optional feature descriptions (User Data Header Indicator = 1), then this first byte counts all the bytes in the optional feature descriptions. After this count byte are the optional feature descriptions themselves (Figure 3-3).

Details of SMS-SUBMIT and SMS-DELIVER

Figure 3-3
SMS_SUBMIT without and with optimal feature fields.

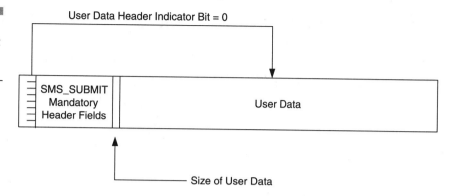

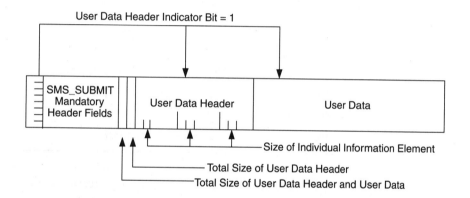

Each optional feature description consists of three subfields:

Subfield Name	Description
Tag	Says which optional feature this is
Length	Number of subsequent bytes
Value	Information describing the optional feature

The tags are, of course, critical. They are carefully controlled by the standards documents and say what the following data is. Here's a list of some of the optional features you can put into the User Data Header in addition to the tag values that identify them.

Tag	Optional Feature
0x00	Reference number for a concatenated short message
0x01	Special SMS message
0x04	Application port addressing
0x06	SMSC control parameter
0x09	Wireless control message protocol parameter
0x0A	Text formatting parameter
0x0B	Predefined sound
0x0C	User-defined sound
0x0D	Predefined animation
0x0E	Large user-defined animation
0x0F	Small user-defined animation
0x10	Large user-defined picture
0x11	Small user-defined picture
0x20	RFC 822 e-mail header
0x70-0x7F	SAT security header

For example, if the message were accompanied by a small animation, then the coding of this optional feature might be:

```
0x0F   0x09   0x01 0x02 0x03 0x04 0x05 0x06 0x07 0x08 0x09
```

Value 0x0F indicates a small animation optional feature, 0x09 indicates that 9 bytes of small animation data follow, and the following 9 bytes are the small animation itself. (It really isn't a real small animation; we are just illustrating the construction of an optional feature description.)

So what are the optional features that might be of interest to application developers?

- **Concatenated Short Messages**—Messages longer than 140 bytes; in fact, up to $255 \times 140 = 35{,}700$ bytes, at least theoretically
- **Special SMS Message Indication**—Message-waiting indicators
- **Application Port Addressing**—Tie into Internet applications
- **USIM Toolkit Security**—Much more about this in later chapters

Details of SMS-SUBMIT and SMS-DELIVER

- **Text Formatting**—Use italics, boldface, and underlining in the message display
- **Sounds to Accompany the Message**—Use unique sounds to announce the arrival of a message
- **Animated Icons to Accompany the Message**—Cute icons announce the arrival of a message
- **Pictures to Accompany the Message**—A picture is worth at least two SMSs

We'll cover each of these eight optional features after we've examined the mandatory Protocol Identifier and Data Coding Scheme fields of the header.

Protocol Identifier

The Protocol Identifier byte tells the mobile receiving the message what to do with it. There are two general possibilities: pass it on to another device or do something with it yourself.

The first possibility is called *telematic interworking* because the mobile serves as an intermediate node between two other devices—FAX machines or pagers or e-mail programs—that are connected via the mobile network. The mobile is interworking telematically with or for these devices and essentially just transports their protocols. If you wanted send a wireless e-mail, for example, you'd set the Protocol Identifier to 0x33, which says that user data part of the message is encoded as an e-mail message and can be passed directly to the local e-mail server.

The second possibility—do something with it yourself—is more interesting. Here are some useful codings of the Protocol Identifier that tell the mobile what to do with the message:

- Value 0x40 is like an Internet ping. It causes the immediate transmission of a message from the mobile phone back to the sender but doesn't log in the message as a received message in the message store of the phone. If you receive this return message, then you have determined that the phone is on without bothering the user.
- Values 0x41 through 0x47 can be used to update status information of some sort. Suppose that you want to build a stock tracking service where you send the current value of a stock to the handset using SMS. With this Protocol Identifier value, you would over-

Protocol Identifier Value	Disposition of the Message
0x40	Acknowledge receipt of message and then throw it away
0x41—0x47	Replace any message with the same Protocol Identifier and the same originating address with this message
0x5F	Enable a return call to the number sending this message
0x74	Reserved for IS-136 R-DATA
0x75	Pass the entire message to the handset's operating system for processing
0x7C	Pass the entire packet to the SIM
0x7D	Pass the entire message to the handset's operating system for processing
0x7E	Pass the payload to the handset's operating system, where it will be used to "de-personalize" the handset
0x7F	Pass the entire packet to the SIM (SIM data download)

write the old value with the new one so that the subscriber could always read the latest value when actually accessing the message.

We used the 0x5F coding in Chapter 2. This coding causes the handset to display the originating phone number with the message and lets the recipient quickly set up a voice call to the number with one or two key hits. This is great for a message that might say "Big News. Call. Fred."

- Values 0x7D and 0x7E are for communicating with applications on the handset. An example of a handset application is a WAP browser. Handsets are just now starting to open up generally useful application programming interfaces. This is how you would use SMS to communicate with those applications.

Details of SMS-SUBMIT and SMS-DELIVER

Later chapters focus on 0x7F for beaming SMS messages to applications on the SIM. Coding the Protocol Identifier is how you get those messages there. When a message arrives with this Protocol Identifier value, the handset knows to pass the message to the SIM. The SIM operating system uses other fields in the message, which we will discuss in detail in subsequent chapters, to know what to do with the message once it gets to the SIM.

It is important to appreciate that passing the SMS message to an application or application on the SIM for processing is quite different from simply storing the SMS message on the SIM for later viewing. For storing the message for later viewing, the SIM is simply used as kind of an industrial-strength floppy disk and the only capability of the SIM that is involved is its file system. An SMS message that is passed to the SIM is actively processed by a program running on the SIM. This processing may cause interaction with the subscriber in the form of message-specific menus and requests for keypad input and it may cause reply SMS messages to be sent. These SIM-based SMS processing programs are your SIM applications and the subject of the second part of this book.

Data Coding Scheme

The overloading of the Data Coding Scheme byte is so complex that this one byte has been given its very own standard, 3GPP TS 23.038. It has to be one of the few bytes in all of computing and communication that has its own standard.

Nominally, the Data Coding Scheme tells the receiving application how the message is encoded. Is it compressed or not? Is it plain old ASCII or the ISO and 3G version of Unicode (called UCS2) or is it just raw seething binary data? Obviously, the application has to know how the message has been encoded to even begin to try to make sense of the bits. But the Data Coding Scheme carries other information such as whether a message is waiting at some other node of the communication network and who should receive that message.

Here are some Data Code Scheme encodings that you might use or encounter.

Data Coding Scheme Value	Description
0x10	Message in GSM default alphabet—display immediately ("flash message")
0x11	Message in GSM default alphabet—send acknowledgment
0x12	Message in GSM default alphabet—give it to the SIM
0x13	Message in GSM default alphabet—give it to the terminal
0x14	Message is binary—display immediately
0x15	Message is binary—send acknowledgment
0x16	Message is binary—give it to the SIM
0x17	Message is binary—give it to the terminal
0x18	Message in UCS2 alphabet—display immediately
0x19	Message in UCS2 alphabet—send acknowledgment
0x1A	Message in UCS2 alphabet—give it to the SIM
0x1B	Message in UCS2 alphabet—give it to the terminal
0xD0	Voicemail message waiting
0xD1	Fax message waiting
0xD2	E-mail message waiting
0xF0	Message in GSM default alphabet—display immediately
0xF1	Message in GSM default alphabet—send acknowledgment
0xF2	Message in GSM default alphabet—give it to the SIM
0xF3	Message in GSM default alphabet—give it to the terminal
0xF4	Message is binary—display immediately
0xF5	Message is binary—send acknowledgment
0xF6	Message is binary—give it to the SIM
0xF7	Message is binary—give it to the terminal

Details of SMS-SUBMIT and SMS-DELIVER

Concatenated Short Messages

It is possible to string short messages together in a kind of train to send messages or data records that are longer than 140 bytes. Each short message segment is still limited to 140 bytes with the concatenated messages option, but the receiving application is alerted that the entire message is not contained within one segment but rather within several incoming segments that have to be combined sequentially, i.e., concatenated, to build the complete message.

In my experience, sending time is not linear with the sequence of packets. The longest wait is for the first packet. Once the first packet gets through, the other packets quickly follow, so you can think of a concatenated short message as a kind of not-so-very-short message.

The header is not repeated after the first packet, but the option header that says that the packet is part of a train is always there (it's the connector between the packets), so you don't get to use the entire payload for your data. Table 3-1 shows the maximum amount of data you can send.

TABLE 3-1 Maximum Message Length Using Concatenated Short Messages

Type of Data	Payload Length	Length of Option Header	255 × (Payload Length − Header Length)
Uncompressed GSM 7-bit default alphabet	160 character	7 bytes	255×153 = 39,015 7-bit character
Uncompressed 16-bit UCS2 Unicode alphabet	140/2 character	6/2 character	255×67 = 17,085 16-bit character
Compressed GSM 7-bit default alphabets	140 bytes	6 bytes	255×134 = 34,170 byte
Binary 8-bit data	140 bytes	6 bytes	255×134 = 34,170 bytes

Although the standard is written to allow a train of 255 cars, most networks limit the number of cars to five or so. Nevertheless, obviously you can say more with 800 characters than with just 160.

The coupler that connects the concatenated short message segments together is an information data element at the beginning of the User Data field and it looks like this:

Subfield Name	Field Value	Description
Tag	0x00	This information data element describes a concatenated short message
Length	0x03	Three bytes follow
Reference Number	$0 <= k <= 255$	All the parts of this concatenated message have the same value k
Total Segments	$1 <= n <= 255$	The total number of segments in the entire message
Index of this Segment	$1 <= i <= n$	Indicates sequence of a particular packet

"You've Got Mail"

In this age of messages popping up in all corners of our lives, it is not surprising that SMS offers three different ways to say, "You've got mail!" We've already seen two of them: the Return Call setting of the Protocol Identifier byte and the Message Waiting settings of the Data Coding Scheme byte. The third one is an optional field descriptor, and this one is not only the most detailed but also the one that can become the all-singing-all-dancing-blot-out-the-sun "You've Got Mail" message.

Subfield Name	Value	Description
Tag	0x01	Special SMS Message Indication tag
Length	0x02	Number of bytes that follow
Message Type	k	0×00 = voice 0×01 = FAX 0×02 = e-mail
Message Count	n	There are n messages of this type waiting

Details of SMS-SUBMIT and SMS-DELIVER

You can put more than one of these Special SMS Message Indicators in the User Data Field if you want to tell the user about more than one kind of waiting message. Normally, the message itself says something about how to retrieve the messages ... or is a threat that your mailbox is overflowing and if you don't empty it soon you may miss that all-important message about having the winning lottery number.

Application Port Addressing

Slowly but surely Internet protocols are seeping into the wireless telecommunications network, and we are starting to see connections to them in various 3GPP standards. The following field is one of many examples. This field has 8-bit and 16-bit versions. The 8-bit version looks like this:

Subfield Name	Value	Description
Tag	0x04	8-bit port tag
Length	0x02	Number of bytes that follow
Destination Port	i	Port to which message should be given
Originator Port	j	Port from which the message came

The 16-bit version is very similar, but the destination and originator port fields have 16 rather than 8 bits.

Subfield Name	Value	Description
Tag	0x05	16-bit port tag
Length	0x04	Number of bytes that follow
Destination Port	i	Port to which message should be given
Originator Port	j	Port from which the message came

The vision is that the handset or the SIM whips up a TCP or UDP packet, drops in these destination and originator port addresses, and sends it off to the SMSC, from where it is delivered to the corresponding application out on the Internet. Setting the port to 0x8000 for example, would deliver the packet to a Web server; setting it to 0x5190 would send the message to an AOL Instant Message server.

SIM Toolkit Security

We will have quite a bit to say about this optional feature descriptor in Chapter 9, where we talk about building end-to-end security. Basically, it is a way for you to encrypt the message.

You probably have heard that all GSM communications are encrypted, so why would anybody want to encrypt a second time? It's true that the network operator encrypts traffic on the wireless connection between the base station and the mobile phone, but SMS messages are clear on other links inside the network, e.g., between the SMSC and the base station. Further, the operator isn't responsible for the confidentiality or integrity of messages before they reach him or after he turns them over to where they are going. In the case of the simple SMS messaging we described in Chapter 1, this wasn't an issue because both endpoints—the sending phone and the receiving phone—and everything in between were inside the network and thus participated in the encryption provided by the network operator.

However, when the mobile phone hands the message to a Web server, which in turn makes it into an e-mail to send to the SMS messaging-enabled application, you want to achieve end-to-end security. This means that the keys are shared between the two endpoints and everybody in between is clueless. SIM Toolkit Security is how the two endpoints can realize end-to-end security. This will be described in detail in Chapter 9.

Enhanced Messaging Services

Enhanced messaging services is fancy talk for fun with your phone. We can use bold text, add sounds, and even send pictures. That's the good news. The bad news is that not all of these features are supported by all handsets, that some handsets have their own proprietary ways, and, worst of all, some network operators won't let you do this stuff at all until they figure out a way to charge extra for it. The situation is slowly getting better. The Multimedia Messaging Service standards effort in

Details of SMS-SUBMIT and SMS-DELIVER

3GPP and the M-Services initiative of the GSM Association seek to create a more standardized environment for enhanced messaging services.

Special effects with text are achieved by including one or more text formatting descriptors. Each descriptor looks like this:

Subfield Name	Value	Description
Tag	0x0A	Text formatting tag
Length	0x03	Three bytes follow
Start Character	n	Character position at which to start the special effect
Number of Characters	k	The number of characters that should get the special treatment
Format Byte		The special formatting to be applied

The format byte is built up according to the following table:

7	6	5	4	3	2	1	0	Description
0								Strikethrough OFF
1								Strikethrough ON
	0							Underline OFF
	1							Underline ON
		0						Italic OFF
		1						Italic ON
			0					Bold OFF
			1					Bold ON
				0	0			Normal sized characters
				0	1			Large characters
				1	0			Small characters
						0	0	Left aligned
						0	1	Center aligned
						1	0	Right aligned
						1	1	Language-dependent alignment (default)

If you set the format byte to 0x14, then the characters from character n to character $n + k - 1$ would be displayed as large boldfaced characters, e.g., "Call me NOW!" A small blessing is that blinking is not supported.

Sounds, Pictures, and Animations

The sounds and pictures that accompany SMS messages should not be confused with ringtones, operator logos, wake-up logos, group graphics, and other such goodies that personalize the handset and its use in plain old voice communication. These features can be and usually are delivered to and installed on your phone using SMS, but they are embedded in special SMS messages that are handled by the handset (see Protocol Identifier = 0x7D). These handset personalization sound and graphic features also tend to be proprietary to particular handsets and particular network operators and sometimes require SMSC support for special features such as the DHMI flag.

The sounds and pictures we are talking about don't get installed on the handset but rather are associated with (or are!) a particular SMS message; think of them as sound and picture accompaniment to SMS messages. For example, suppose you wanted to send a message "Dinner moved to Sally's house" and wanted to accompany the message with a little map of how to get to Sally's house from where the dinner was previously scheduled to be. You'd accompany the message with a little picture that the recipient could pop up on the screen to get to Sally's house.

Unlike ringtones and wake-up logos, which are installed semipermanently on the handset, when the SMS message containing sounds or pictures is deleted, the accompanying sounds and pictures go away. Also, unlike ringtones and logos, no special permission or support from the SMSC is needed to send those items to a phone. However, support by the handset is required and, as we have observed, not all handsets support those enhanced messaging services.

There are 10 predefined sounds that you can call forth from the handset using the following optional feature descriptor:

Details of SMS-SUBMIT and SMS-DELIVER

Subfield Name	Value	Description
Tag	0x0B	Predefined sound tag
Length	0x01	1 byte follows
Value	n	Index of predefined sounds

The predefined sounds that are available are:

Predefined Sound Index	Description
0	Chimes high
1	Chimes low
2	Ding
3	Ta da
4	Notify
5	Drum
6	Claps
7	Fanfare
8	Chord high
9	Chord low

There's something very charming about "ta da" showing up in an international telephony standard, isn't there?

You can also generate your own sound by using the iMelody format and then send it to the phone in an SMS message with the following optional feature descriptor:

Subfield Name	Value	Description
Tag	0x0C	User defined sound tag
Length	n	n bytes follow
Value		n bytes of iMelody format sound

iMelody is a simple and elegant format for melodies of single notes. It was created by the Infrared Data Association. Details of the iMelody format are available at their Web site, www.irda.com. A note is an optional octave prefix, followed by the note, the duration, and an optional duration specifier, all in the default GSM alphabet. Thus, middle C sharp held for two beats would be *3#c2. This might be a little verbose for SMS messaging but doubtless could be the source of endless phun (fun with your phone).

Unlike sounds, the 3G standards don't predefine pictures (although it's fun to ponder what they'd be if they did!) but you can build your own pictures in one of three sizes: small, large, and variable.

Small pictures use the following optional feature descriptor:

Subfield Name	Value	Description
Tag	0x11	Small picture tag
Length	0x20	32 bytes follow
Value		Bytes of the picture

Large ones use the following optional feature descriptor:

Subfield Name	Value	Description
Tag	0x10	Large picture tag
Length	0x80	128 bytes follow
Value		Bytes of the picture

Pictures are encoded into the bytes that represent them, starting from the upper left hand corner and then moving from left to right across each row, until the lower right hand corner is reached. Each byte encodes eight pixels, with bit number 7 being the leftmost pixel. There are no gray scales or colors; 1 means a black pixel and 0 means a white pixel.

As I'm sure you've noticed, screen resolutions on mobile phones aren't great but are surprisingly useful for line-drawing communications graphics such as maps. The following table gives some examples of what's presently available.

Details of SMS-SUBMIT and SMS-DELIVER

Handset	H × V (pixels)
Sanyo SCP-6000	120 × 96
Nokia 8890	84 × 48
Alcatel 301	96 × 38
Motorola Timeport	96 × 40
Siemens S35	101 × 80
Philips TCD 998	81 × 57

The bad news is that they are all different, which makes crafting a picture that doesn't care about what phone it is sent to a challenge, to say the least. Unlike the SIM, which is highly standardized and thus presents an attractive application platform, handsets are still in the pre-PC phase of their evolution, where gratuitous differences are touted as features.

Animations, like sounds, offer some predefined ones and let you do it yourself if you don't like what's built-in. The predefined animations are accessed with the following optional feature descriptor:

Subfield Name	Value	Description
Tag	0x0D	Predefined animation tag
Length	0x01	1 byte follows
Value	k	Index of predefined animation

The animations that can be summoned forth are:

Predefined Animation Index	Description
0	I am ironic, flirty
1	I am glad
2	I am sceptic (sic)
3	I am sad
4	WOW!
5	I am crying

And I'll bet you thought the European Telecommunications Standards Institute didn't have a sense of humor.

Do-it-yourself animations are encoded as four sequential pictures. There are two sizes of do-it-yourself animation: large, which consists of four 32 × 32 pictures, and small, which consists of four 16 × 16 pictures. Large animations are encoded in the following optional field descriptor:

Subfield Name	Value	Description
Tag	0x0E	Large animation tag
Length	0x80	128 bytes follow
Value		Four 32 × 32 pictures

Small animations are encoded in the following optional field descriptor:

Subfield Name	Value	Description
Tag	0x0F	Small animation tag
Length	0x20	32 bytes follow
Value		Four 16 × 16 pictures

Internet E-Mail

Many people think that getting e-mail on a mobile phone is the killer application for wireless data. The last optional feature descriptor we consider describes an Internet e-mail in the message field. The descriptor is disarmingly straight forward given the complexity we've seen for doing much simpler tasks, such as stating the time of day:

Subfield Name	Value	Description
Tag	0x20	Internet e-mail tag
Length	0x01	1 byte follows
Value	k	RFC 822 e-mail header length

Typically this will be the only (or at least the last) option feature descriptor in the User Data field, so the next *k* bytes make up the RFC 822 header for the e-mail and the remaining bytes in this SMS segment and any concatenated segments are the e-mail message itself. The hope is that the recipient's handset includes software that can grok RFC 822 headers and will present the e-mail to the user in some in some familiar format. If not, the handset will dump the text, RFC 822 header and all, into the incoming SMS message file and the recipient will have to parse RFC 822.

SMS_DELIVER

SMS_SUBMIT is all about sending the SMS message. SMS_DELIVER is about what arrives at the other end. The surprising thing is what you send isn't exactly what your correspondent receives. Everything in the User Data field arrives exactly as it was sent, complete with all the optional feature descriptors and all the User Data, but the header is different (Figure 3-4).

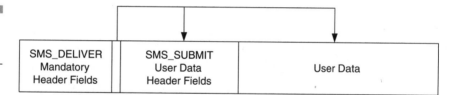

Figure 3-4
SMS_DELIVER headers.

The mandatory SMS_DELIVER header consists of 19 bytes including the 8-byte originating telephone number. The 11 nonaddress bytes are:

Field Name	Bytes	Description
SMS Flags	1	General information about the packet
Protocol Identifier	1	How the SMS packet is going to be used
Data Coding Scheme	1	How the message is encoded
Service Center Timestamp	7	Time that the SMSC received the message
User Data Length	1	Length of the following User Data field

The 1-byte SMS Flags field in SMS_DELIVER consists of six subfields:

Subfield Name	Bits	Description
Message Type Indicator	2	0x00 for SMS-DELIVER
More Messages	1	If set to 0, there are more messages waiting at the SMSC for this mobile; if not, set to 1
Reply Path	1	If set to 1, the receiving mobile should allow easy selection of sending the phone number to return a message or initiate a voice call
User Data Header Indicator	1	There are optional feature descriptions after the mandatory header and before the message itself
Status Report Indication	1	If set to 1, the sender has requested a delivery status report

The Service Center Timestamp is 7 bytes long and nibble swapped. *Nibble swapped* means that the low-order 4 bits of the byte come first and then high-order 4 bits come second in the received bit stream. Using the message we received in Chapter 2:

	Year	Month	Day	Hour	Minute	Second	Time Zone
Nibble-swapped digits	0x10	0x20	0x11	0x80	0x23	0x44	0x58

The message becomes:

```
2001 February 11, 08:32:44, GMT-5
```

The hours are coded on the 24-hour clock. By nibble swapping the time zone, we get 0x85 because bit 7 is 1, a negative number. The 3APP TS 23.040 standard says that the −5 is the number of quarters of an hour between Boston time and GMT, which is wrong because then

Details of SMS-SUBMIT and SMS-DELIVER

the entry would be 0x94. This is a problem for my GSM operator, VoiceStream, and their SMSC provider. Your GSM operator undoubtedly sticks by the standard.

Although the network won't (or at least shouldn't) delete any of the optional feature descriptors in the User Data field, it might add some. In that case, a special separator descriptor is used to mark where the descriptors authored by the original sender end and where the descriptors inserted by the network begin. This separator descriptor is called the UDH Source Indicator and looks like this:

Subfield Name	Value	Description
Tag	0x07	UDH Source Indicator tag
Length	0x01	1 byte follows
Value	0x03	Following descriptors created by the SMSC

When this descriptor appears, the network has inserted all the following descriptors up to the beginning of the message data itself. If the original message with the original optional feature descriptors just fit into one SMS segments, the single SMS might turn into a multisegment concatenated SMS because adding the network option fields makes the total message longer than 140 bytes. You might get grumpy about this if the operator charges you for two SMS messages rather than one because of the added stuff.

Summary

This chapter describes in detail the construction of the delivered SMS message—SMS_SUBMIT—and the received SMS message—SMS_DELIVER. By toggling various bits in the outgoing message, you can tease a lot of useful behavior out of the SMS and send a lot of useful information through to the mobile. Aside from the USIM Toolkit Security Headers in a later chapter, we will consider SMS messages more holistically as simply carriers of our messages. Nevertheless, it is good to know that you can open the hood and tinker with the system to get an effect that isn't available when we regard short messages as simply pneumatic tubes that carry bits.

CHAPTER 4

SMS Integration

We've taken a look under the hood of SMS messaging and discovered a lot of capability there. Sending an SMS alert with a sound and an embedded telephone number for quick response was an example of a useful SMS application that could be run by itself or easily added to an existing application. In this chapter we consider the tighter coupling of SMS messaging to applications and using it to create a mobile extension of those applications rather than just a fire-and-forget alert mechanism.

AT commands are fine for sending messages but a bit awkward for receiving them into an application because you have to continually poll the handset to see if anything new has arrived. Fortunately there are a number of commercial products that not only present a high-level interface to sending messages but also provide some programmer-friendly constructs to make receiving SMS messages a lot easier.

Nokia makes a package called the Nokia PC Connectivity Software Developers Kit (SDK) that is available for free download from www.forums.nokia.com. It is long on programming flourishes but short on functionality and, of course, only works with Nokia handsets. GPA Technology in Australia sells a little package called SMS Gateway that is very useful. You can download a trial version from www.winsms.com. There's another nice package called SMS-IT available from www.sms-it.com. For the truly adventurous there is the Kannel Open Source WAP and SMS Gateway available from www.kannel.org.

We're going to use a package from Derdack Software Engineering that is available at their Web site, www.derdack.com. The package is called the Message Master Developer Suite. I picked this package because it is reasonably priced, offers high- and low-level access to the phone, is simple and straightforward to use, and the folks at Derdack were very responsive to my questions.

All these packages work on basically the same principle. There are two queues, incoming and outgoing, of messages between your application and the phone and the network behind it. Rather than dealing directly with the modem, your application puts messages into the outgoing queue and takes messages off the incoming queue. You can set a number of parameters of the queues including the frequency with which the queues interact with the modem to send messages out or gather new messages in and what functions should be called when various modem events (such as the arrival of a message) occur (Figure 4-1).

SMS Integration

Figure 4-1
Application connecting to Message Master SDK.

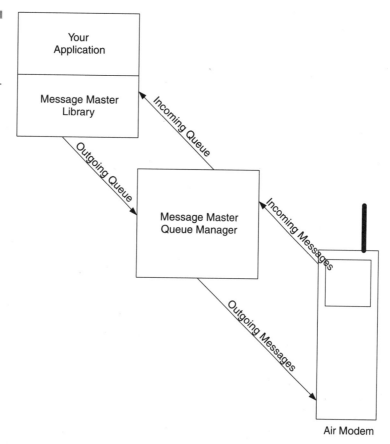

We'll begin our discussion of SMS integration with an example. Although the primary purpose of the example is to illustrate how to work with a high-level SMS package, the example itself—connecting a database to mobile telephones—is not wholly contrived and might be something you want to do.

OK, suppose we have a database—in our case, it's an Excel spreadsheet but it could be any database—that we want to query from a mobile phone. The database queries and the database responses will use SMS (Figure 4-2).

For the sake of the example, let's say that our particular database contains the list of a corporate sales team's active customer contacts. The contacts could have come in through the company's Web site, they could have come in through the company's call center, or they could have been e-mailed in. However they came in, they are written

Figure 4-2
Mobile access to an SQL database.

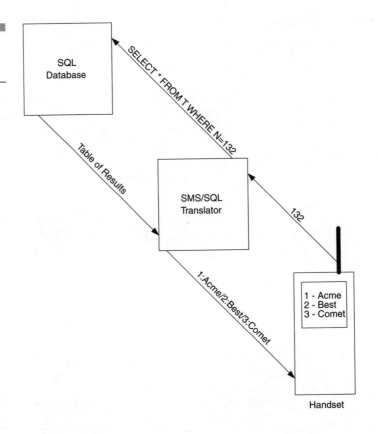

into the customer contact database and are available for access by any member of the sales team. Our task is to make this database available on the team's mobile telephones.

The each record in the contact database has six fields:

RecordNo—the sequential number of the record in the database

Company—the customer company from which the contact came

AccntRep—the last four digits of the mobile phone for the account representative for this company

Number—the return call telephone number

Person—the name of the customer that initiated the contact

This last field says why the customer contacted and can contain one of four values:

Subject—a categorization of the nature of the contact

Ordr—I wish to place an order

SMS Integration

Info—I want to be sent some information
Trbl—I'm having trouble with one of your products
Meet—I'd like to schedule a meeting with you

A traveling account representative starts a session with the database simply by sending an SMS with the single digit 0 in the body of the message. When this SMS is received, the last four digits of the phone it came from are used in a SELECT command to find the records in the database that belong to this account representative. An abbreviated list of active records that are the responsibility of the account representative is then sent back. For example, the rep might get the message:

1.Comet(Trbl)
2.Best(Meet)
3.Acme(Info)
4.Best(Ordr)

This would mean that there are four customer contacts in the database for that representative; somebody at Comet is reporting a problem, somebody at Best wants to schedule a meeting, somebody at Acme wants some information, and somebody at Best, perhaps somebody else wants to place an order.

To find out more about one of those active contacts, the account representative simply sends back the number of the contact that they'd like to find out move about. In the case at hand, the representative is worried about the trouble report and so sends back a 1.

Back from the database comes an SMS message with a sketch of the selected record:

Comet
1.Numb(415)
2.Pers(Sara)
3.Subj(Trbl)

This may be enough information for the representative to give Sara a call or she may want to go ahead and retrieve the phone number. A reply of 1 will do that:

Sarah Jones
Number
415 446 5678

Now, just by selecting Use Number on the handset, the representative can call Sara and find out what the trouble is.

Message Master lets us build this interactive application by providing a function named MMIRegisterCallbacks that lets us identify functions that are to be called when various events happen on the modem:

```
MMIRegisterCallbacks(MessageInfo, MessageSent,
MessageReceived, &lResult);
```

The MessageInfo function is called when Message Master wants to give some information to the user. MessageSent is called whenever a message is actually sent to the SMSC. MessageReceived is called whenever a message is received. Of course, you'd call MMIRegisterCallbacks with your own functions:

```
extern long  infoFn(), sentFn(), recvFn();
MMIRegisterCallbacks(infoFn, sentFn, recvFn, &lResult);
```

Message Master calls your MessageReceived function with five arguments:

```
MessageReceived(long lMessageType, LPCTSTR lpszOriginator,
LPCTSTR lpszDestinator, DATE dateTimeStamp, LPCTSTR
lpszMessageString, LPCTSTR lpszInfo)
```

lMessageType indicates the message's origin. In our case it will always be 1 indicating that it is coming from SMS. lpszOriginator is a string that holds the telephone number of the mobile sending the SMS. lpszDestinator is the phone number of the mobile receiving the message. Message Master allows you to receive messages on more than one serial line and, hence, more than one handset, so it's important to know on which handset the SMS has arrived. dateTimeStamp is the SMSC timestamp that we discussed in the SMS_DELIVER section of Chapter 3. lpszMessageString is the message itself, which in our case will be numbers if it is coming from one of our mobile account representatives. lpszInfo currently is not used by the package.

The first thing our MessageReceived function does is pick off all the digits in the incoming message:

```
/*
** Pick up all the entered digits
*/
keyhits = sscanf(lpszMessageString, "%1d%1d%1d%1d%1d",
 &keyhit[0], &keyhit[1], &keyhit[2], &keyhit[3],
 &keyhit[4]);
```

If there are none, then we don't know who the message came from, so we drop it on the floor:

```
/*
** If the message doesn't start with a number it is not
   for us
*/
if((keyhits == 0) || (keyhits == EOF))
  return 0;
```

Using the telephone number for authentication isn't exactly public key cryptography but it does offer some protection. If the sender had to enter a personal identification number (PIN) to activate the phone, then we would have the same two-factor level of security as an ATM card: you have to have something, the SIM, and you have to know something, the PIN, in order to query your database. If it's good enough security for banks to dispense money, then it is reasonable to use this level of two-factor security for a database query. Other levels can be added as we discuss in Chapter 9.

Because a database query can involve the exchange of a number of SMS messages with the mobile, when we get a message we have to determine whether this message is part of an ongoing exchange or the first message of a new exchange.

Don't forget that, we are supporting many account representatives in the field simultaneously, so that the messages arriving from them could be intermixed. For example, the first message from one account representative could be followed by the third message in an ongoing exchange with a second account representative and that by the second message in the conversation with a third account representative.

Therefore, the next patch of code determines whether we are in the middle of an exchange with the telephone number in the incoming message and, if so, retrieves the current state of that exchange:

```
/*
** See if there is an open session for this phone number
*/
k = -1;
for(i = 0; i < CONTEXTS; i++) {
    if(mContext[i].mState == 0) {
        k = k == -1 ? i : k; // save the open session
        continue;
    } else if(strcmp(lpszOriginator,
mContext[i].mNumber) == 0) {
        k = i;
        break;
    }
}
```

If this is the first time we've seen this account representative, we open a context for them:

```
/*
** If this phone number doesn't have an active session,
** start one
*/
if(mContext[k].mState == 0) {
  mContext[k].mNumber = _strdup(lpszOriginator);
  mContext[k].mState = 0;
}
```

Now we process the numbers in the SMS message one at a time and compose our response SMS message in lpszResponse. Notice that we have allowed multiple key hits. This is for account representatives that know what information they want and don't want to walk through all the intermediate menus. The program behaves just as if it had sent out the intermediate menus and processes the key hits as if they were responses to those (unsent) intermediate menus.

```
/*
** Loop over all the key hits
*/
for (key = 0; key < keyhits; key++) {

    strcpy(lpszResponse, "");
```

SMS Integration

```
/*
** If keyhit is 9, then go to the top
*/
if(keyhit[key] == 9) {
  mContext[k].mState = 0;
}

/*
** If keyhit is 0 then go "up" one menu level.
** If this is the last key hit then return the
** results from that menu level by setting the
** key hit to the last key hit we saw at this level.
** Otherwise, go to the next key hit.
*/
if(keyhit[key] == 0) {
  mContext[k].mState = mContext[k].mState > 1 ?
                                   mContext[k].mState-1 : 1;
  pick = mContext[k].mKeyhit[mContext[k].mState];
} else {
  mContext[k].mState++;
  pick = keyhit[key];
}

/*
** Save the current keyhit at this level
*/
  mContext[k].mKeyhit[mContext[k].mState] = pick;

switch(mContext[k].mState) {
```

The states which a conversation with a particular account representative can be in are the following:

```
/*
** States
**
**   0 Empty Context - new conversation
**   1 Pick Context - display contacts - pick one contact
     to examine
**   2 Default Data Context - return default information
```

```
       and fields
**  3 Particular Data Context - return specified data
       field
*/
```

In state 1 we simply collect the database records associated with the last four digits of the telephone number in the incoming SMS and compose a little menu of those records to send back. The challenge is to make optimum use of the screen on the mobile. You want to abbreviate everything so that you can get as much as possible on the screen but you don't want to be so brief that the account representative can't figure out what is going on. What we've done here is to clip each field and value to the first four characters.

```
case 1: /* Return active contacts list */

  rs = new CRecordset(&m_db);

  sprintf(filter, "AccntRep=%s",
          &lpszOriginator[strlen(lpszOriginator)-4]);
  rs->m_strFilter = filter;
  rs->Open(AFX_DB_USE_DEFAULT_TYPE,
"SELECT * FROM Contacts", CRecordset::none);

  i = 1;
  while(!rs->IsEOF()) {
    rs->GetFieldValue("Company", company);
    rs->GetFieldValue("Subject", subject);
    sprintf(item, "%d.%4.4s(%4.4s)\n",
                               i++, company, subject);
    strcat(lpszResponse, item);
    rs->MoveNext();
  }

  mContext[k].mRecords = i-1;

  rs->~CRecordset();

  break;
```

In state 2 we send back a sketch of the selected record:

SMS Integration

```c
case 2: /* Make chosen message the current record. */
    /* Return default data and field list. */

  rs = new CRecordset(&m_db);

  sprintf(filter, "AccntRep=%s",
          &lpszOriginator[strlen(lpszOriginator)-4]);
  rs->m_strFilter = filter;
  rs->Open(CRecordset::snapshot,
          "SELECT * FROM Contacts ", CRecordset::none);

  i = 1;
  while(!rs->IsEOF()) {
    if(i++ == pick)
      break;
    rs->MoveNext();
  }

  if(rs->IsEOF()) {
    sprintf(lpszResponse, "Only %d messages", i-1);
    return 1;
  }

  for(i = 0;
         i < sizeof(currentRecord)/
            (2*sizeof(currentRecord[0][0]));
         i++) {
    free(currentRecord[i][1]);
    rs->GetFieldValue(currentRecord[i][0], field);
    currentRecord[i][1] = _strdup(field);
    if(i == 0) {
      sprintf(lpszResponse, "%s\n",
                              currentRecord[i][1]);
    } else {
      sprintf(item, "%d.%4.4s(%4.4s)\n",
                     i, currentRecord[i][0],
                              currentRecord[i][1]);
      strcat(lpszResponse, item);
    }
  }
}
```

```
    rs->~CRecordset();

  break;
```

In state 3 we send back the selected field of the selected record:

```
case 3: /* Return chosen data field for current record */

  if(pick > mContext[k].mRecords) {
    sprintf(lpszResponse, "Only %d messages",
                          mContext[k].mRecords);
    return 1;
  }

  sprintf(lpszResponse, "%s\n%s\n%s",
                        currentRecord[2][1],
    currentRecord[pick][0],
                        currentRecord[pick][1]);

  break;
```

Here's the patch of code that sends out the response that we composed above. We just use the number that came in lpszOriginator as the number to which we send the response:

```
if(ComposeResponse(lpszOriginator, lpszMessage,
lpszResponse)) {
  if((MMIMessageSend(&lIndex, lpszService,
      lpszOriginator,
    "", time.operator DATE(), 0, lpszResponse)) != 0) {
    AfxMessageBox("Error sending reply.");
  } else {
    lpszNote.Format("Msg sent to %s", lpszOriginator);
    ::SetDlgItemText((
(CDialog*) theApp.m_pMainWnd)->GetSafeHwnd(),
                    IDC_SENTMESSAGES, lpszNote);
  }
} else {
  if((MMIMessageSend(&lIndex, lpszService,
lpszOriginator,
    "", time.operator DATE(), 0, "No such entry")) != 0) {
```

SMS Integration

```
            AfxMessageBox("Error sending reply.");
        } else {
            lpszNote.Format("Error sending reply.",
                                          lpszOriginator);
            ::SetDlgItemText(
                ((CDialog*) theApp.m_pMainWnd)->GetSafeHwnd(),
                                IDC_SENTMESSAGES, lpszNote);
        }
    }
```

Figure 4-3 shows the state transition diagram of how we handle the interaction with each account representative. Simply stated, each state refines the context of the previous state and shows more information about fewer items until we get to the end, where we show the complete information about a single item. At this point the caller can go

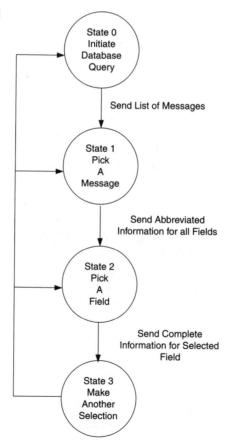

Figure 4-3
User interaction state diagram.

back and broaden the scope—less information about more items—and take the refinement dialog down another path.

Summary

There are many ways to improve this little application. For example, you probably want to provide some way for the account representative to indicate that a message has been handled so that it doesn't keep appearing on the list of active messages. You also might want to proactively send an SMS message with a sound alert to an account representative if a contact came in from their biggest account or when any new message for that representative came in at all.

What we have achieved with surprisingly few lines of code are:

- Adding mobile access and query capabilities to an SQL database using SMS
- Handling SMS messages from multiple mobiles simultaneously
- Maintenance of the state of the conversation with each mobile
- Crafting of context-sensitive responses that balance information content, screen size, and interaction

What made this so easy is that the Derdack SMS Systems Developer Kit (SDK) handled all the interactions with the handset and the network and simply called our application when a message arrived. Further, when it did, it had dissected the SMS_DELIVER header and passed us useful bits and pieces of the header in a useful format in addition to the message itself.

As with all high-level programming interfaces, however, what we gain in ease of use we lose in control over details. In Chapters 2 and 3 we wrestled with every bit and byte of an SMS message. This let us control everything that was available to control, but it was a very code-intensive and error-prone environment. If we were one bit off, the message didn't go through.

In this chapter, we gave up some of that control to get on top of a more facile programming interface. Our application came up quickly, and the SDK took care of almost all of the details of creating and disassembling SMS messages.

In the next and last chapter on SMS, we explore an even higher interface to SMS messaging: SMS brokers.

CHAPTER 5

SMS Brokers

Using a GSM phone as an air modem is great if the amount of SMS traffic you are sending or receiving is small, but the bandwidth quickly will become a bone in the throat as you leave the prototype stage of your development and ramp up to commercial use. Fortunately, there are higher-bandwidth connections to the SMS system available.

Many GSM operators offer direct connections to the short message centers. These can be dial-up, X.25, or Internet connections. Many different protocols are used on these connections, and the majority of these are proprietary to the company that sold the SMSC to the operator. Examples are:

SMSC Vendor	SMSC Connection Protocol
ADC NewNet	SMCI (Short Message Client Interface)
CMG	UCP (Universal Computer Protocol)
CMG	EMI (External Machine Interface)
Comverse	ISMSC (Intelligent Short Message Service Center)
Ericsson	CAP II (Computer Access Protocol #2)
Logica	SMPP (Short Message Peer to Peer)
Motorola	CDMP (Cellular Digital Messaging Protocol)
Nokia	CIMD (Computer Interface to Message Distribution)
SEMA	OIS (Open Interface Specification)
SEMA	SMS2000

SMPP has some traction as a common protocol (Figure 5-1) and one organization, the SMPP Forum (www.smpp.org) promotes it.

Even if you get a thick pipe to the network operator, you are faced with the operator's SMS delivery and reception policies. Unlike voice, where there is universal roaming, SMS connectivity between operators is uneven at best. An operator may deliver SMS messages to some operators and not to others. An operator also may be willing to receive SMS messages from some operators and not from others. Finding out who is connected to whom can be a daunting task. Much of this confusion is due to the fact that operators haven't figured out how to charge each other for carrying each other's SMS traffic.

Where there is a demand and a reluctance to satisfy it, there is a business opportunity, so into the breech has stepped a new commer-

SMS Brokers

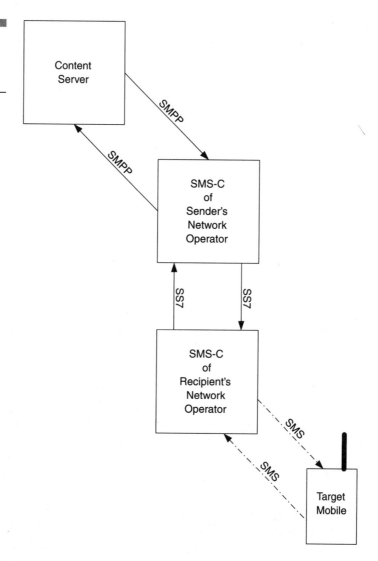

Figure 5-1
Message flow using direct connection to an SMS-C.

cial entity called the *SMS broker.* An SMS broker is kind of a virtual network operator for short message traffic. They strike deals with all the operators, get the thick pipe connections, and provide an interface between your SMS traffic and the operator who can deliver it to where it is going. One of the important advantages of using SMS brokers is that most of them deliver messages not only to GSM phones but also to Time Division Multiple Access (TDMA) phones, Code Division Multiple Access (CDMA) phones, pagers, and PDAs such as Palm Pilots and Visors.

The cost of sending and receiving SMS messages with an SMS broker depends mostly on your monthly volume. Prices start at around 10 cents per message for volumes in the hundreds and thousands and drop to 2 to 3 cents per message for volumes in the millions.

Some SMS brokers limit their services to specific countries. Others will deliver a message anywhere in the world. Examples of global SMS brokers are:

Broker Name	Web Site
Annyway	www.annyway.com
CMG	www.cmg.com
Diax	www.diax.ch
Dr. Materna	www.materna.com
Inphomatch	www.inphomatch.com
NovelSoft	www.sms-wap.com
Quios	www.quios.com
wapMX.com	wapmx.com
Winbox.com	www.winbox.com

SMS brokers offer different ways of giving them a message to send and different ways of getting a message from a mobile back to you. Sending is a lot easier than receiving, so let's start with that. Of course, before you send or receive, you have to have an account with the SMS broker so that he can bill you for your activity (Figure 5-2).

Perhaps the simplest way of sending an SMS message through a broker is with an HTTP POST to the broker's server. The POST can be from an executable program written in your favorite programming language such as C, Java, or Visual Basic, or it can be from a Web page—server side or client side—that uses your favorite scripting language such as VBScript, JScript, or Perl.

The fields included in the POST vary from broker to broker but at a minimum the fields will include your account number, the telephone number that is to receive the message, and the message itself. As with the high-level programming interfaces discussed in Chapter 4, there aren't many SMS features that you can use through a POST interface. Mostly it's limited to sending text messages, although some brokers do support pictures and sounds.

SMS Brokers

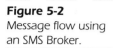

Figure 5-2
Message flow using an SMS Broker.

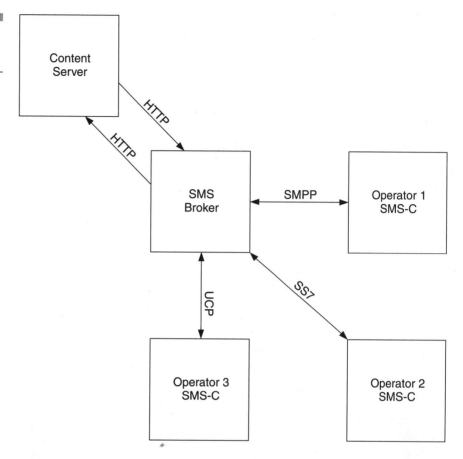

Here's what sending a SMS message through the NovelSoft service looks like as rendered in Visual C++ using the CinternetSession class:

```
LPCTSTR FastSMSHeaders =
"Host: www.sms-wap.com\nContent-type: application/x-www-form-\
urlencoded\nContent-length: ";

LPCTSTR FastSMSRequest =
"UID=xxxxxxxx&PW=xxxxxxxx&N=+16172301346&M=Hello,     world\n";

  m_pInetSession = new CInternetSession(_T(    "Mobile-Mind"), 1,
                        INTERNET_OPEN_TYPE_DIRECT);
```

```
m_pHttpSession = m_pInetSession->GetHttpConnection(
        "www.sms-wap.com", 80, "xxxxxxxx", "xxxxxxxx");

m_pHttpFile = m_pHttpSession->OpenRequest(
    CHttpConnection::HTTP_VERB_POST, "/cgi/csend.cgi");

strcpy(Headers, FastSMSHeaders);
sprintf(RequestLength, "%d\n", strlen(FastSMSRequest));

strcat(Headers, RequestLength);

strcpy(Request, FastSMSRequest);

try {
   m_pHttpFile->SendRequest(Headers, strlen(Headers),
                    (void *)Request, strlen(Request));
}
```

This code sends the message "Hello, world" to my mobile.

Here is how the same sending message would look as rendered in a client-side VBScript:

```
<HTML>
<HEAD>
<TITLE>Send Message Immediately</TITLE>
</HEAD>
<BODY topmargin=10>
Sending an SMS Message to +1 617 230 1346
<SCRIPT LANGUAGE=VBScript>

Dim HttpObj          'HTTP Obj
Dim HttpUrl          'HTTP URL for NovelSoft account message page
Dim UID              'User ID
Dim PW               'Password
Dim N                'Telephone number of mobile
Dim M                'Message content

' Instantiate the HTTP COM object
Set HttpObj = CreateObject("AspHTTP.Conn")

' Message sending parameters
```

SMS Brokers

```
UID     = "xxxxxxxx"
PW      = "xxxxxxxx"
N       = "16172301346"
M       = "Hello, world"

HTTPObj.Url = "http://www.sms-wap.com/cgi/csend.cgi"

HTTPObj.PostData = "UID=" & HttpObj.URLEncode(UID) & "&" & _
            "PW="   & HttpObj.URLEncode(PW)  & "&" & _
            "N="    & HttpObj.URLEncode(N)   & "&" & _
            "M="    & HttpObj.URLEncode(M)

HTTPObj.RequestMethod = "POST"

' Send the message sending request to the server
HTTPObj.GetURL

</SCRIPT>
</BODY>
</HTML>
```

The same message sent as a server-side ASP would be coded the following way:

```
<% Language = VBScript

Response.write("Sending an SMS Message to +1 617 230 1346")

Dim HttpObj         'HTTP Obj
Dim HttpUrl         'HTTP URL for NovelSoft account message page
Dim UID             'User ID
Dim PW              'Password
Dim N               'Telephone number of mobile
Dim M               'Message content

' Instantiate the HTTP COM object
Set HttpObj = CreateObject("AspHTTP.Conn")

' Message sending parameters
UID     = "xxxxxxxx"
PW      = "xxxxxxxx"
```

```
N       = "16172301346"
M       = "Hello, world"

HTTPObj.Url = "http://www.sms-wap.com/cgi/csend.cgi"

HTTPObj.PostData = "UID=" & HttpObj.URLEncode(UID) & "&" & _
           "PW=" & HttpObj.URLEncode(PW) & "&" & _
           "N="  & HttpObj.URLEncode(N)  & "&" & _
           "M="  & HttpObj.URLEncode(M)

HTTPObj.RequestMethod = "POST"

' Send the message sending request to the server
HTTPObj.GetURL
%>
```

This is all rather straightforward Web programming. Things get a little more challenging when you want to get out of couch-potato mode and interact with the mobile. The one-way mode can blast out alerts, horoscopes, and the joke of the day, but is doesn't allow the recipient to reply. (Of course, for the joke of the day, you might not want to answer!) The real value in SMS messaging is two-way communication, and doing this by way of an SMS broker requires some additional moving parts.

Because you are really using the broker's GSM account and telephone number, any reply to messages sent through them obviously will come back to that account. If the broker has more customers than just you—and you certainly hope he does because you and the other customers share the cost of running the brokerage—then there has to be a way for the broker to figure out which of his customers should receive the incoming SMS reply.

Now the broker could keep track of who sent the outgoing message so that, when the reply came back, he'd just send it to you, but doing so would require a massive database and this approach breaks down if two or more people are using the broker to send messages to the same mobile.

NovelSoft has set up a clever service that lets you register one or more keywords that are unique to you. When a message comes in from a mobile that starts with one of your keywords, they send a GET containing the message to a Web server that you've associated with the keyword. You process the message on your Web server, and whatever you respond to NovelSoft's GET they send back to the mobile.

For example, I have registered the keyword MM (for mobile mind) with NovelSoft. If I send you a message out through NovelSoft, it will arrive at your mobile as if it came from their mobile phone, which in the spring of 2001 had the phone number +41 4002030. If you now respond to this message but start your message with the letters MM, it will go via the GSM SMS back to +41 4002030, where NovelSoft will see the MM and send a GET of the message along with your mobile number to http://www.mobile-mind.com/mm.asp. Presto, chango, we are in two-way communication.

Of course, this works the same way if you rather than I initiate the conversation. That is, if you send an SMS message out of the blue to +41 4002030 that starts with MM, then NovelSoft is going to send the GET, and we're off to the races.

To illustrate how this all works, let's build a little m-commerce server. This is going to be a state-of-the-art, multistep application like the database application we used in Chapter 4. In this setup, your customers have been preidentified and have established credit with you. They have chosen PINs and told you the telephone number from which they will be placing their orders. Further, the orders are from a small, well-defined, and static catalog so customers only have to provide the catalog numbers of the products they want.

This setup is perfect, for example, for placing standing refill orders for one of a small number of standard items: "Send me my usual reorder of Lava Soap." No need for your customer to chat on the phone or interact with your Web page. Three quick SMS messages while waiting at the traffic light and they've placed the order directly into your order entry system and driven on. This order process is illustrated in Figure 5-3.

The customer starts the conversation by sending in his or her PIN. This establishes a connection between (knowledge of) the PIN and the phone sending it and thus authenticates the customer. You actually could consider eliminating this step and just authenticate the customer with the phone number. Adding the PIN provides greater security.

The start message looks like this:

```
MM PIN 1234
```

After checking to make sure the PIN goes with phone number of the sender, our little m-commerce server responds with:

```
BUY OR CHK?
```

Figure 5-3
Message flow for the reorder application.

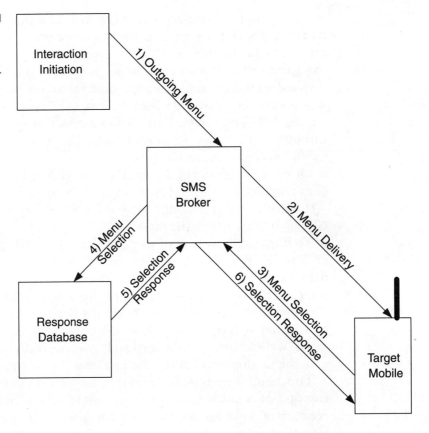

This asks the customer whether he or she wants to place a new order or check on the status of an existing order.

To check on the status of an existing order with the order number 567, the customer would simply respond with:

```
MM CHK 567
```

To enter an order for catalog item number 89, the customer would respond with:

```
MM BUY 89
```

Remember that the MM is how NovelSoft is routing the message back to the application server.

In the first case, our server might respond with:

```
CHK 567 SHIPPED
```

In the second case, the server might respond with

```
BUY 89 568
```

where the 568 is the order number the system assigned to that order.

Here are the SMS-relevant parts of a server-side ASP script that implements our m-commerce server:

```
<% Language = VBScript

Dim HttpObj     ' HTTP Obj
Dim HttpUrl     ' URL for transaction engine

Dim mob         ' Mobile telephone number
Dim msg         ' SMS message
Dim trn         ' Transaction record
Dim tok         ' Tokens of the message
Dim rpy         ' Reply to the mobile phone
Dim ord         ' Order number
Dim sta         ' Status of an order

mob = Request.QueryString("ORIG")
msg = Request.QueryString("MSG")

tok = Split(msg)

If(tok(1) = "PIN") Then

' Associate the PIN with the mobile phone number to
' authenticate the user and open a conversation

rpy = "BUY OR CHK?"

ElseIf tok(1) = "BUY" Then

' Make sure you have a conversation going with this
  phone
```

```
' Enter the order for item number tok(2) and put
' the order number in the variable ord

rpy = "BUY " & tok(2) & ord

ElseIf tok(1) = "CHK" Then

' Make sure you have a conversation going with this
  phone

' Look up the status of the order whose order number
' is in tok(2) and put it in the variable sta

rpy = tok(2) & sta

Else

rpy = "ERROR"

End If

'
' Log the customer interaction into the CRM system
'
Set HttpObj = CreateObject("AspHTTP.Conn")

trn = mob & "|" & msg & "|" & rpy

HTTPObj.Url = "http://www.crm.acme.com/mobile.dll"

HTTPObj.PostData = "Transaction=" &
HttpObj.URLEncode(trn)

HTTPObj.RequestMethod = "POST"

HTTPObj.GetURL

'
' Send the reply back to the phone
'
```

```
Response.ContentType = "text/plain"

Response.Write mob & "," & rpy

end if

%>
```

Now you definitely shouldn't try this at work. A real m-commerce server would worry a lot more about authentication and authorization, as you will see in the SmartSignature case study in Chapter 14. The example simply illustrates how easy it is to connect to an SMS broker.

The only place where this approach gets into choppy water is in the nonuniform nature of SMS connectivity around the world. As you probably observed, the phone number you had to send your order to was in Switzerland. If you were in Switzerland, this would work just fine. However, if you happened to be in say, Boston, then it wouldn't work because VoiceStream, a Boston GSM provider, doesn't forward messages to Swisscom, the Swiss provider. It must be said that this is a recognized problem and mobile network operators of all technologies around the world are working feverishly on getting data coverage as seamless and universal as voice coverage.

One solution to this problem is to find a broker connected to VoiceStream. Then you can catch messages coming from Boston, but you may be out of luck if one of your customers travels to Switzerland.

As we observed, every service not offered by the network operators is a business opportunity for somebody else, so the brokers have come up with a solution for this one too. Get a VoiceStream account and put it in Switzerland. You can do this because your account is in the SIM and you can pop the SIM into an overnight FedEx to Switzerland. This approach to the problem is called SIM hosting.

Now your customers enter the VoiceStream telephone number rather than NovelSoft's Swisscom number. VoiceStream recognizes the account as one of their own and sends the message. Of course, "you" happen to be in Switzerland (or wherever your SIM host is). NovelSoft catches the message off your VoiceStream SIM just as they did off their own and handles the message in the same way.

SIM hosting can be a little expensive—$300 per month on top of NovelSoft's SMS charges plus VoiceStream's SMS roaming charges—but it does solve the international connectivity problem.

Summary

We discussed using SMS brokers to build SMS applications and provided a couple of examples. The advantages of using SMS brokers are:

- International coverage
- Uniform interface to many different kinds of mobile devices
- Easy to program
- Internet connectivity
- Message tracking
- Scaling to high volumes

The disadvantages are:

- Additional cost
- Lack of access to all features of SMS messaging

If the application you are extending to the mobile network is already Web based, like the model illustrated in Figure 5-4, or if the folks you want to connect to are international, then working with an SMS broker certainly should be considered. Even if the application is local but high volume, it may be easier to work with an SMS broker than with the local network operator.

SMS is the ugly ducking of mobile networks. It was conceived as a way to squeeze a couple extra pennies out of some arcane underutilized bandwidth and exploded into an extremely popular system feature. More than 20 billion SMS messages are sent globally every month.

There are those who say that, as advanced digital services such as GPRS arrive on the scene, SMS will ride off into the sunset. It is the view of others including ourselves that transactions are the definitional heart of mobile communications behavior. Therefore, although it may change its name, SMS is a mobile messaging paradigm that is here to stay.

Quite unlike SMS, which was not initially envisioned to be a major source of income, the SIM was developed explicitly to enable network operators to create new revenue streams. However, unlike SMS, it has taken a long time to generate widespread availability of SIM-based applications. Nonetheless, all of the necessary building blocks are in place.

The SIM is a secure, tamper-resistant computer inside the 3G handset that is owned and controlled by the network operator. It sports an

SMS Brokers

API and can host multiple applications. After a look at a live SMS application, we'll explore how to build applications for the SIM. First we will look in more detail at how an SMS-based application can create real value for a corporate customer.

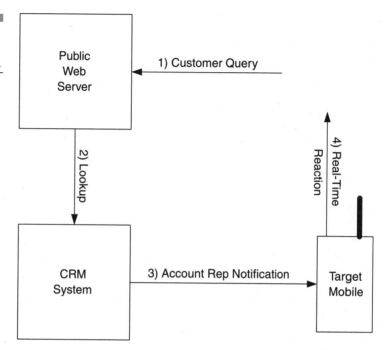

Figure 5-4
SMS adds real-time response to the Web.

CHAPTER 6

SMS in an Airport Logistics Application

SMS Case Study: Atraxis

It is easy to recognize the limitations of simple text messaging on the small display screen of a typical second-generation GSM mobile phone. These limitations have not stopped millions of users around the world from enthusiastically embracing SMS text messaging. These wireless customers are not waiting for third-generation capabilities to turn the phone into a device that handles data as readily as voice. With the number of SMS messages exchanged each year projected to exceed 200 billion, sending data over mobile phones has become a worldwide cultural phenomenon.

The most frequently publicized SMS applications are oriented to consumers and not corporations. Is there any role for simple SMS-based solutions in a business-to-business setting? The answers is "yes," as this case study will demonstrate. The widespread usage of SMS in daily life also provides a strong rationale for enterprise adoption. A growing population of employees who are already facile with text messaging on mobile phones essentially eliminates the need for special training. Moreover, the simplicity of menu-driven, single-key responses combined with the cost effectiveness of phones for data input makes for a compelling business case for SMS.

The Atraxis Group, an international provider of logistical IT solutions and consulting services for the aviation industry, saw the very simplicity of SMS messages as an attractive implementation feature. Atraxis serves more than 60 airlines and some 40 airport authorities around the world. Supported by a comprehensive global network of partnerships and joint ventures, the company, headquartered in Zurich, Switzerland, has a workforce of more than 2,100 professionals worldwide.

The managers of Atraxis' Airport Engineering, Consulting, and Innovation division at Zurich's international airport were looking for a fast and cost-effective means of improving the communication between a central, mainframe-based Atraxis program called the axs-Control Departure Control System and the various crews preparing a plane for departure. This effort was part of a broader initiative aimed at generally increasing efficiency and punctuality at the Zurich airport. The overarching goal in Zurich was to make sure that more planes took off on schedule without compromising quality or safety in any way.

The specific objective for this "on the ground" extension of axsControl was to gather messages about the status of each crew's work on an airplane as quickly as possible into the Atraxis Departure Control System, so that the plane could be cleared for takeoff as soon as it was ready. This had to be accomplished at low cost, with a minimum of disruption to existing work routines, and the need for little to no training of work crews in how to use the system. In addition, it was important that the input be fed automatically into axsControl so it was received without the need to use intermediate "man in the loop" data handling or elaborate and expensive user interfaces.

The timeline for coming up with a mobile input solution for axsControl was extremely tight—the overall project to improve punctuality at the Zurich airport was scheduled to begin in spring 2001. That meant whatever approach Atraxis selected for implementing the mobile extension to axsControl, it had to go from design to testing to full implementation in just a few months. The solution that the Atraxis team designed was a combination of simple phone menus based on SIM Toolkit and SMS messaging. It is called SMS2axsControl.

Project Background

SMS2axsControl is part of a new system Atraxis has implemented at Zurich airport. It consists of a Web-based application, the Hubwatch, and the axsControl system, which collects all the status messages related to a particular aircraft preparation for takeoff. axsControl is Atraxis' departure control system that governs the airlines' core processes of passenger, baggage, and aircraft handling. It is a hosted application that runs on the mainframe system in Zurich and serves about 40 customers at more than 150 airports worldwide.

Under the name "Hubwatch," axsControl offers a way to control the various services to be performed during the ground-stop of a flight that are related to the actual arrival and departure timings. This includes de-boarding of passengers; offloading baggage; cargo and mail; fueling; cleaning; catering; getting the crew on board; loading; boarding; and pushback of the aircraft.

The start and termination of the various tasks are entered, either automatically or manually, into the axsControl system by the various service providers. Manual inputs are done directly on the dedicated axsControl terminals or via Internet/Intranet screens. It is also planned to link resource management systems to Hubwatch.

Various service providers such as fuel companies, cleaning staff, cabin crew, and catering companies involved in the preparation of an aircraft for takeoff are expected to inform the axsControl about the current status of the work being performed. Service providers that work from a single location have a dedicated terminal linked to axsControl to update the readiness database. At the start of the project, however, there was no easy, fast way to collect real-time information from service providers with workers who did not have access to a terminal. That meant that the information in the axsControl database could lag behind the actual completion of a given task. If the crew responsible for fueling the plane was late in delivering its input about the status of that process, for example, the entire departure countdown could be delayed.

It was clear that such a tool could be operated much more successfully if the staff assigned to the various jobs could update the system completely and immediately. For that reason, it was obvious that mobile input units would make the task easier.

Atraxis and Zurich airport managers believed that making a mobile data input device available to all of the service crews in the field working on the planes would make the whole turnaround process more efficient. Once each field crew was equipped with a mobile phone capable of real-time interactions with the axsControl database, the flow of communication would be continuous as long as the input process was simple, fast, and accurate.

Focus on the Essentials

The two most critical scheduling points in an airplane-servicing task are the beginning and the end. So the SMS2axsControl focused on these two essential information points. Whenever a task was started or completed, the field crew sent a formatted SMS to a predefined telephone number. The SMS indicated which crew was sending the message, whether the task was starting or finishing, and the serviced flight number. This simple SMS message was then forwarded to the axsControl system, which automatically updated scheduling routines that monitored all the activities related to a given flight.

The main reasons for selecting an SMS-based solution were the cost effectiveness and ease of implementation offered by SMS on a mobile phone. Atraxis wanted to use low-priced, widely available equipment to keep down the cost of the application being deployed in a variety of airport situations. Because the entire project had to be completed in two months, it was not realistic to consider solutions that required any hardware development or relied on devices that were not available off the shelf. The whole point was to contribute to the Zurich airport punctuality program, so there was considerable pressure for a fast development and adoption cycle to meet a targeted launch deadline of April 2001.

In terms of the overall project implementation and long-term maintenance cost, Atraxis managers didn't want to rely on voice messages or simply calling in the status of a job on a mobile phone because this would require a more expensive support and input structure at the central database. Voice input would require staff who would be receiving the messages and entering the data into axsControl.

The use of phone with SMS input also seemed a good way to reduce the need for training different service providers' staff members. Because most workers would already have personal mobile phones with SMS capabilities, they would be familiar with the text input process and using a menu display on the phone. That meant that, as long as the application was simple and intuitive, Atraxis did not have to include any formal training program in its implementation plans.

Design and Development Process

The primary developers for the SMS2axsControl SIM program were two engineers from the Atraxis division that handles airport logistics in Zurich. Their role was to design the overall communication flow, including the menu structure, and provide an interface between the SMS messages arriving from SMS2axsControl and the axsControl Departure Control System. That interface was called the axsControl Bridge.

SMS2axsControl uses SIM cards that have been specially configured with the menus that the field crews use to input their status information. Atraxis worked with Swisscom, the local GSM network operator, to create these SIMs and a developer from Swisscom wrote a SIM

Toolkit application for the SIM that displayed the appropriate menus and sent the input information out as an SMS message.

Getting the application designed and implemented on the SIM with the collaboration of Swisscom was only the first component in getting SMS2axsControl up and running. The second component was moving those SMS messages over the air from the mobile phone to the axsControl system. To help with that part of the project, Atraxis enlisted AnnyWay, which operates one of the largest European SMS service centers. The third component of the communication flow was a connection between the SMS service center and the Internet. This function was handled by the AnnyWay Information Center, which is located in Dortmund, Germany. An overiew of the system design and communication flow of SMS2axsControl is shown in Figure 6-1.

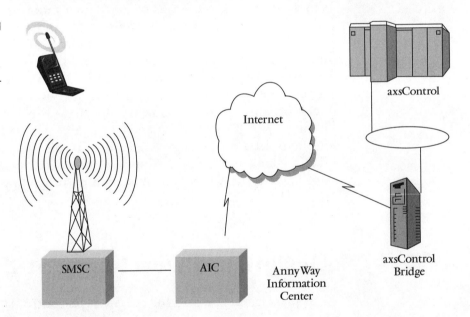

Figure 6-1
SMS2axsControl system overview.

The overall message flow is as follows:

- The user sends a formatted SMS message via mobile phone to Swisscom SMS-C.
- The SMS message is forwarded to the AnnyWay Information Center (AIC)
- The AIC contacts a Web server on the axsControl Bridge and passes on the SMS message together with origin phone number

SMS in an Airport Logistics Application

- A servlet is started on the axsControl Bridge, which converts the formatted SMS message into an axsControl message that is then sent to the axsControl mainframe.

Details of the axsControl Bridge components and architecture are shown in Figure 6-2.

Figure 6-2
The axsControl Bridge architecture.

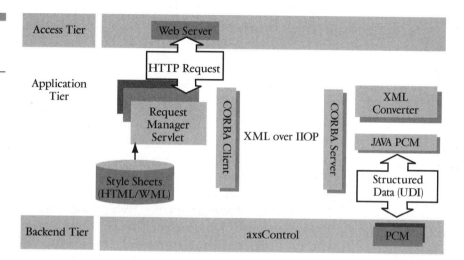

The Action on the Ground

The menu structure on the mobile phone was designed by Atraxis and Swisscom to be as intuitive as possible. It was deliberately kept very simple so that almost all of the input is based on a dichotomous (either/or) menu pick. The only text input is the number of the flight being serviced. At the same time, the menus offer flexibility and a variety of options to fit different situations that need to be reported to the axsControl Departure Control System.

Figure 6-3 shows the menu structure used by the fueling crew. Because many airlines might be using the application, selecting the appropriate airline and inputting the flight number is a part of the process.

Figure 6-3
Menu for the fueling crew.

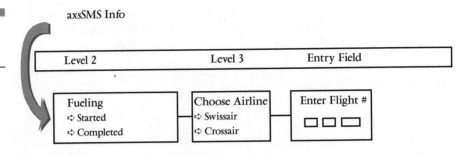

As shown in Figure 6-3, the fueling crew member picks axsSMS Info on the main menu on the mobile phone and is presented with the two-item pick menu: Started or Completed. After one item has been chosen, the second two-item menu containing the names of the two airlines subscribing to the service, Swissair and Crossair, immediately appears. As soon as the user selects the appropriate airline, the third and final screen appears, which asks the fueling crew member to key in the flight number that is being serviced. The whole string is then presented to the user for verification. When the user selects OK, the SMS message is sent automatically to a predefined telephone number.

The menu structure used by the cockpit and cabin crew is a little more complicated, as shown in Figure 6-4. Typically, the crew inside the plane share one phone to report on the plane's status, which is the reason the cabin and cockpit menus appear on phones with this SIM.

Figure 6-4
Menus for the cabin crew.

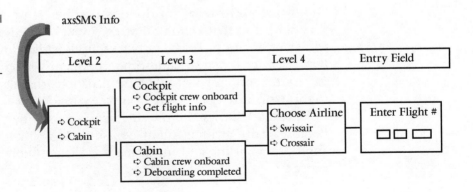

As Figure 6-4 shows, the menu for the cockpit crew includes an option to search for additional flight information. This function was developed as a secondary application called SMS4Pilots. If the pilot needs flight-related information of the next flight or leg, he or she sends a formatted SMS message to a predefined short number. The SMS message is forwarded to the axsControl Departure Control System. axsControl collects the information from different systems and returns an SMS message containing all the requested information.

Project Performance Review

Within two months of its launch in April 2001, about 200 phones with SMS2axsControl were being used in production mode at the Zurich airport. The two largest carriers at the airport, Swissair and Crossair, had signed on to use the applications to expedite communication with the axsControl Departure Control System.

Using an SMS interface on the phone lived up to its promise of being very efficient in terms of the deployment and adoption processes. There was no problem in getting the different types of users comfortable with the system. Atraxis prepared a short presentation to explain the purpose of the SMS2axsControl application, but no in-depth training or orientation was needed. As expected, the airport staff was used to sending SMS messages, so this system wasn't difficult to adopt.

All the basic departure readiness input is menu driven with very simple information displays and selections. These menus are designed to fit on today's small-screen phones. The airlines using the application today are running it on Motorola Timeports.

Atraxis provided phones and SIMs to get the airlines started. They selected Motorola Timeport due to its triple-band capabilities, because the flight crews also wanted to have mobile phones available at destination airports around the world to use for voice communication. This has proved to be a very popular choice with the users. The mobile phone equipment is already very familiar, and the crew wants to carry a phone anyway so no extra device is required. The fueling crew uses Ericsson R310s, which is better suited for outdoor use.

The only training that was required was an introductory presentation about SMS2axsControl. There was no need for specific training in each function because the menus are self-explanatory. This is an important user advantage because many staff members rotate on plane crews all the time.

The only significant user complaint concerns occasional lapses in SMS performance. Performance of SMS in general is not always as good as it needs to be for a time-sensitive application, which means that SMS reliability and response time can present a problem from the user's point of view and for the application as a whole.

In designing the project, developers estimated that the minimum effective response time would be 2 minutes or less. Currently the typical response time to get a message through the system is 30 seconds. Ninety percent of the time it is 2 minutes or less, which is acceptable, but there is also an issue with reliability of SMS because it is not run as a business-level service with a guarantee of delivery. Too much interruption in SMS availability is a challenge that Atraxis has to address before this application can be extended to airports in other parts of the world. They are working with AnnyWay and Swisscom to improve performance and reliability.

Evaluating the Business Results

The direct customer for SMS2axsControl is the airline. So far, two airlines Swissair and Crossair, have signed up for the service in Zurich. Swissair served as the field test, with an initial deployment of 80 phones. As of June 2001 the application had expanded to more than 200 phones being used by both airlines and was being used to expedite approximately 400 departures each day.

The airlines pay a fee to Atraxis for the SMS2axsControl setup and application implementation and a surcharge for each SMS message sent through the system. The costs associated with implementation are relatively low because no special mobile equipment is involved and the number of messages sent during the departure process for each flight is limited. Therefore, cost is not seen as a barrier to more airlines adopting this application.

How about results? Is there evidence that real time mobile communication to the departure control system is helping to keep departing planes on schedule? Even after the first few months of implementation, some improvement in the efficiency of the departure process and the number of on-time departures was noted. Because the application has been in use only for a short time, detailed evaluation hasn't been conducted. Atraxis expects to measure departure improvement for participating flights over time and compare performance of planes using the SMS2axsControl application with those without it.

The business benefits for Atraxis are significant. It can provide a valued service to its airline customers and help them meet their overall goals for improving their on-time departure records. SMS2axsControl also has the internal benefit of further automating the departure control system to accept incoming SMS messages. Expansion to other airlines and sale of the application to other airports are being planned. At the same time, Atraxis is working with Swisscom and AnnyWay to address reliability of SMS response time.

Summary

Simple SMS messages, particularly when augmented with task-specific menus, can be used effectively in commercial and business-to-business settings. The costs of the handsets and the messaging benefit from the enormous economies of scale provided by widespread consumer use. This widespread use also virtually eliminates the need for training.

Commercial applications that might consider SMS as the communication medium are those that move small amounts of data on time scales measured in minutes. The quality of service levels of all SMS systems is of concern, but the value yielded by those applications is sufficiently large to be easily shared with network operators in exchange for better service.

CHAPTER 7

The SIM

When you light up your mobile phone, the first question that occurs to the mobile network operator isn't "Who would you like to call?" but rather "Who is going to pay for this call?" Before GSM, the answer was simple. The operator retrieved an account number from the phone, looked it up in his records, and if the account was in good standing, he knew who was paying for the call and the conversation could continue. If the caller was not a customer of that operator, then obviously the account would not be found and the conversation stopped.

One of the primary driving forces behind GSM was using a mobile phone on somebody else's network, which is called *roaming*. GSM developed in Europe, where each country had its own mobile phone company but people continuously roamed between countries. The need for business harmonization—"you can use my network even though you are a subscriber from another country"—led to technical harmonization. GSM is primarily thought of as a technical wonder, which it certainly is, but it is also very much a business model wonder.

So you visit somebody else's network and light up your phone there. That same old question comes to the fore: "Who is going to pay for this call?" Obviously, the network you are visiting doesn't have an account for you so they can't answer the question the old way. What they can do is retrieve from your phone the identity of the network that does hold your accounts and ask them whether you're good for your charges. The visited operator sends your home operator your account number and gets back a "yes, we'll pay this guy's bill" or a "no, we wouldn't let him call his dying mother." Then, when you're done talking, the visiting operator sends your home operator a bill for your call. Your home operator pays the bill and adds it to your monthly charges (with a modest handling fee, of course).

This would work swimmingly if nobody ever tried to make calls they didn't want to pay for. In particular, your home operator wants to make sure it's really you out there and not just somebody that has some how stolen your account number. Obviously, if you got a bill that said you made calls in three different countries at about the same time, you would have a pretty strong case that at least two of them weren't really you.

A simple and ingenious solution to this problem is at the core of the GSM network. Put one of exactly two copies of a secret cryptographic key in your phone and put the other copy in your account back at home. When you light up while roaming, the visited network retrieves your account number from your phone and sends it to your home net-

The SIM

work. The home network looks up your account and, if it is in order, sends back a random number and an encryption of that random number with the key in your account: "I'm good for the charges if the guy you are talking to can encrypt this random number and get this result." The visited network sends the random number to your phone and compares what it gets back with what it got from your home network. If the two numbers are identical, you can start dialing for dollars. This process is illustrated in Figure 7-1.

Of course, this whole scheme turns on there only being two copies of the secret key. If a third copy came into being, then you theoretically could be in two countries at the same time. This is where the SIM comes in. Remember that SIM stands for Subscriber Identity Module. It is very literally the keeper of one of the two keys.

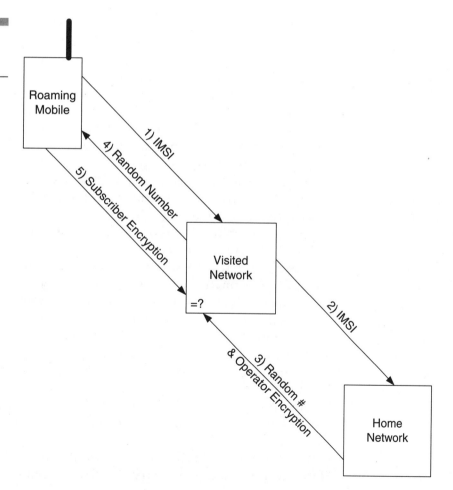

Figure 7-1
GSM authentication protocol.

The GSM engineers needed a secure place to store the subscriber's copy of the key to make their scheme work but this secure place also had to be able to perform some computations with the key to encrypt that random number. It would make no sense to fret about storing it securely and then pass it to the handset when it went into action. By happy coincidence, there was just such a device lying about—the smart card. It could store data securely and perform computations with that data. Just shave off all the card's decorative but useless plastic, pop it in the phone, load up a key, and we're literally on the air. And that is all a SIM is—a garden-variety smart card without the plastic with a secret key inside. Poke a random number in and get an encrypted version out. The key never leaves its secure package, and it's almost impossible to tear the package apart to get the key out. Perfecto!

The network operator owns the SIM just like the credit card company owns the credit card. It is, after all, the operator's on-the-scene agent who is vouching for you and underwriting your use of somebody else's network. If the two numbers match, then your network operator is on the hook for your use of the foreign network.

The choice of the smart card as the keeper of the secret key ended up having a nice side effect. You could move it from handset to handset. In this way, when a new handset came out that you just had to have, you could move your account to it yourself without bugging the operator to move your keys about. Anybody who has tried to change his or her handset on a non-GSM network will appreciate this feature. Not only is it a hassle for the subscriber, it is a lost cause expense for the operator.

All of this engineering took place at the dawn of the development of the GSM network in the mid to late 1980s. Once it was decided to use a smart card for the SIM (there were actually three competing proposals but the smart card won out), a number of additional uses for this tamper-resistant mobile computer were found.

For example, your real account number isn't really sent over the air back to your home network. Rather, a temporary account number is used that can be changed regularly by your home network. This prevents capturing your account number and using it to track where you are. Your current temporary account number is stored on the SIM just like the secret key. The only difference is that the temporary account number can be and is changed periodically, whereas the secret key isn't.

Another example is the use of the computational capabilities of the SIM is to generate a temporary key that is used to encrypt the communication between the handset and the network. This is the air link and the link that is most subject to freeform eavesdropping. After you have been successfully authenticated, the SIM uses the random number to create a key that is used to encrypt the traffic between the phone and the visiting network. This key also is generated by your home network and given to the visiting network over landlines so that the key used for protecting the phone and the network never travels over the air.

There are many books and papers on the security and cryptography used in the GSM network, so if this aspect of GSM strikes your fancy, there is much to learn. We've included some references to this body of literature in the bibliography if you want to pursue this path. For the purpose of application development, what is important is that the SIM is a secure computer that is underpinned with perhaps the biggest key infrastracture outside the government. The SIM and this key infrastructure can be harnessed to inject trust into your applications and the transactions they generate.

SIM-based applications aren't about a daily horoscope or a late-breaking football score. We'll leave those to untrustworthy handset applications based on WAP or whatever replaces it. SIM-based applications are about taking action with the assurance that the data on which the action is based is from who is says it is from (authenticated and nonreputable), that it hasn't been tampered with (integrity), and that nobody else is privy to what is happening (confidentiality).

I like to think of my GSM phone as the ultimate remote control from which I can issue commands about every aspect of my life from anywhere in the world to anywhere in the world. This makes the SIM a very, very attractive application development platform.

Smart Cards 101

Just as there is a large body of information on GSM cryptography, there are many sources of information about smart cards. The cryptography we can assume because it really happens under the covers. To use the SIM, we have to understand a little about smart cards because, at the most fundamental technical level, a SIM is a just a smart card that happens to have a mobile phone as its reader.

A smart card contains a hierarchical file system just like your desktop computer or laptop computer. Rather than having human readable names like "My Documents" or "Chapter6.doc," directives and files in a smart card consist of two binary bytes. Thus we speak of the directory 0x7F01 or the file 0x45C6. The root of the hierarchy is called the Master File (MF). Its name is always 0x3F00.

There are different kinds of files, just like on your PC, but not nearly so many different kinds. There are directories, of course, and files that contain an unstructured block of data such as a name or an account number. There are also record files that contain a series of fixed-size records, each of which may have well-defined, fixed-size fields. Directories are referred to as DFs (unless they contain all the information pertinent to a specific application, in which case they are called ADFs) and files are generically referred to as EFs (Figure 7-2).

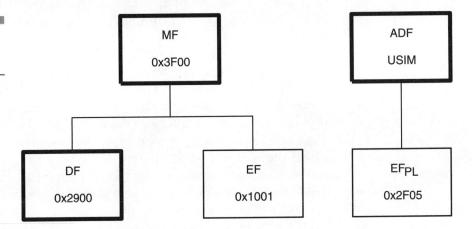

Figure 7-2
Examples of directories and files.

There are about 80 files on a SIM card. Because not everybody can remember that 0x2F05 is the file that contains the subscriber's preferred language, the convention has been adopted to use a descriptive subscript on DF or EF when talking or writing about these files. The language preference file, for example, would be referred to as EF_{PL} (for Preferred Language). Nevertheless, when you talk to the SIM about this file, you have to use its two-byte name, 0x2F05.

Talking to a smart card such as a SIM means sending it commands to perform actions. For example, if we wanted to retrieve from the SIM the subscriber's language preference, we would use the SELECT FILE command to focus the SIM's attention on EF_{PL} and then the READ BINARY command to instruct the SIM to return to us the contents of the file.

The SIM

> If the SIM is just a smart card, why can't we use it in a plain old smart card reader? Why do we have to get at it through the handset or even through the entire GSM network? The fact is you can communicate with a SIM through a plain old smart card reader, but you can do only those things that you the subscriber are allowed to do. However, this way of using the SIM can be an advantage. In fact, it can produce some very interesting applications.
>
> An even more exciting possibility is using the SIM as a way to inject trust into transactions coming from your laptop or desktop computer. After all, it is still you, whether it is you standing at a bus stop with a mobile phone in your hand or sitting at your desk wanting to buy a book on Amazon.com. At this point however, many network operators are so busy trying to control all the applications on the mobile network that they don't seem to understand that they can offer additional trusted services and make money in other networks such as the very nonmobile World Wide Web.

Let's get down and dirty and take a look at the actual bytes that are sent to the SIM to read the subscriber's language preference. First, we have to focus the SIM's attention on the file from which we want to read data, namely good old 0x2F05. To do this we send the SELECT FILE command to the SIM. SELECT FILE for 0x2F05 looks like this:

CLA	INS	P1	P2	Lc	File ID	
0xA0	0xA4	0x00	0x60	0x02	0x2F	0x05

It's seven bytes long and consists of six fields. The CLA ("class") field provides some information about the format of the command. Value 0xA0 means this is a GSM command.

The INS ("instruction") value 0xA4 says that this is the SELECT FILE command. P1 ("parameter 1") isn't used in this command and the P2 (not surprisingly, "parameter 2") value 0x60 says don't send back any information about the file, just focus all of your attention on it. The Lc value indicates that the next file name is 2 bytes long. The File ID bytes are the name of the file. Lots of other options are available by putting codes in the other fields, but we'll concentrate on the simple case for now.

Even though we said don't send back any information about the file, the SIM always sends back a 2-byte status word saying how things went. If the 2 bytes, 0x90 and 0x00, come back, then everything went OK and the SIM is ready for the next command. If something went wrong, then the SIM would send back an error code. For example, if it couldn't find the file you asked for, it would send back 0x6A 0x82, which translates freely as "File not found."

OK, we got back a 0x9000 so the SIM is concentrating on 0x2F05, the preferred language file, EF_{PL}. The subscriber's languages are listed in this file in order of preference, with the most preferred language being first. Each language is coded on two bytes as two alpha-numeric characters. The coding of languages as two characters is given in ISO 639. Let's read the first 2 bytes using the READ BINARY command:

CLA	INS	P1	P2	Le
0x00	0xB0	0x00	0x00	0x02

The INS value of 0xB0 says that this is the READ BINARY command. P1 and P2 together are offset into the file where the read should being. In our case, because we want to read the first 2 bytes, the offset is 0. The Le value indicates how many bytes we want to read.

In this case, the SIM sends back 4 bytes:

```
0x65 0x6F 0x90 0x00
```

The first 2 bytes are the encoding of the subscriber's most preferred language and the second 2 bytes are our old friend 0x9000, which from the SIM's point of view indicates that everything went OK. The first 2 bytes are ASCII for "eo," which means the person prefers to speak Esperanto.

The commands sent to the SIM to ask it to do something are called APDUs. APDU stands for Application Protocol Data Unit. The number of commands that the SIM understands has grown over the years. Today, there are 22 APDUs that the SIM understands. All interaction with the SIM boils down to the handset sending one or another APDU to the SIM and catching what comes back. Even when you put your own application on the SIM, as we discuss in Chapter 12, it will have to define the APDUs that must be sent to the SIM to talk to it.

The Evolution of the SIM

Before we go into details, we should say something about the SIM hardware and software architecture and about how these are evolving.

From its first use in a GSM phone until very recently, a SIM was an 8-bit microcontroller, usually an Intel 8051 or a Motorola 6805. The amount of memory in the SIM has grown. Initially containing only 256 bytes of RAM and 3 kilobytes of EEPROM, it has grown to 1,024 bytes of RAM and 32 kilobytes of EEPROM. All the software for the SIM—the operating system and the code that handles the file system and the APDUs—is burned into 32 to 64 kilobytes of ROM. The 80 odd data files are kept in the EEPROM, and the RAM is used as an I/O buffer for communication with the handset. This is about all that is needed for the SIM to perform its cryptographic duties and provide some relatively trivial telephony services such as storing a phone book of frequently dialed numbers.

Beginning in the mid-1990s, as new applications started to find their way onto the mobile telephone, the SIM started to morph into a full-fledged application platform. Needless to say, this put all kinds of evolutionary pressure on the SIM operating software and the SIM hardware. The software started to sprout virtual machines, principally the Java virtual machine, to host nontelephony applications, and the hardware started to grow additional memory and computational abilities.

Roughly speaking, there have been three generations of SIMs. In the first generation, SIM was just a SIM. The only thing the SIM smart card could do was be a SIM. The smart card hardware was committed solely to SIM functionality. In the second generation, additional applications could be added to the SIM. The SIM was still the uber-application, but it could call on other applicationettes to create custom services for the subscriber. In the third and current generation, the SIM itself becomes an application and all applications have equal (well, almost equal) standing on the platform.

Generation	Name	Characteristic
1	Purpose-built SIM (Figure 7-3)	SIM performs network authentication and universal telephony services such as phone book and SMS message storage
2	SIM with applications (Figure 7-4)	Additional customer-specific services can be added as SIM Toolkit (STK) applications after the SIM is in the field
3	SIM as an application (Figure 7-5)	SIM is one of the multiple telephony applications, specifically the authentication application, on the platform

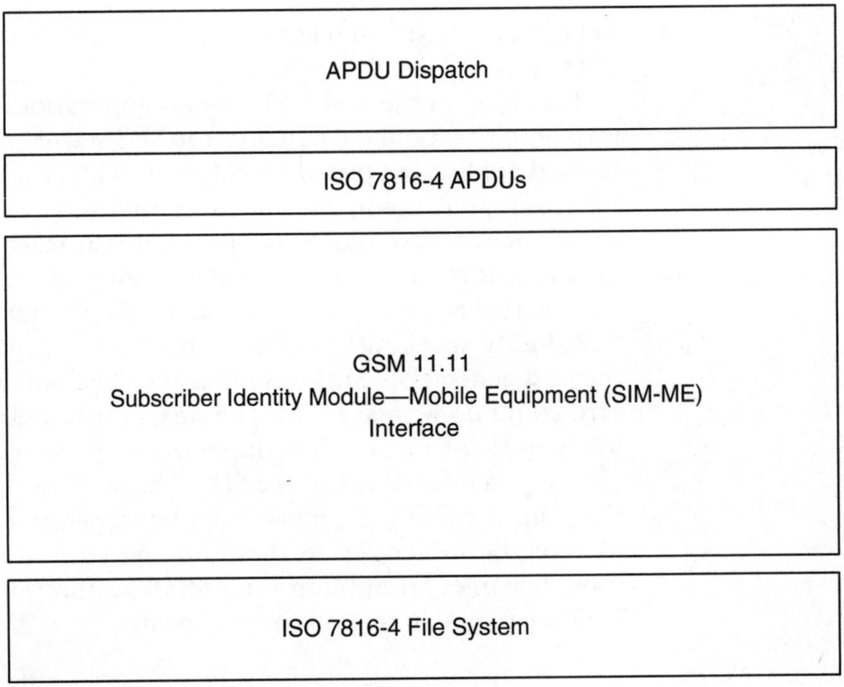

Figure 7-3
Architecture of a first generation SIM.

APDU Dispatch

ISO 7816-4 APDUs

GSM 11.11
Subscriber Identity Module—Mobile Equipment (SIM-ME)
Interface

ISO 7816-4 File System

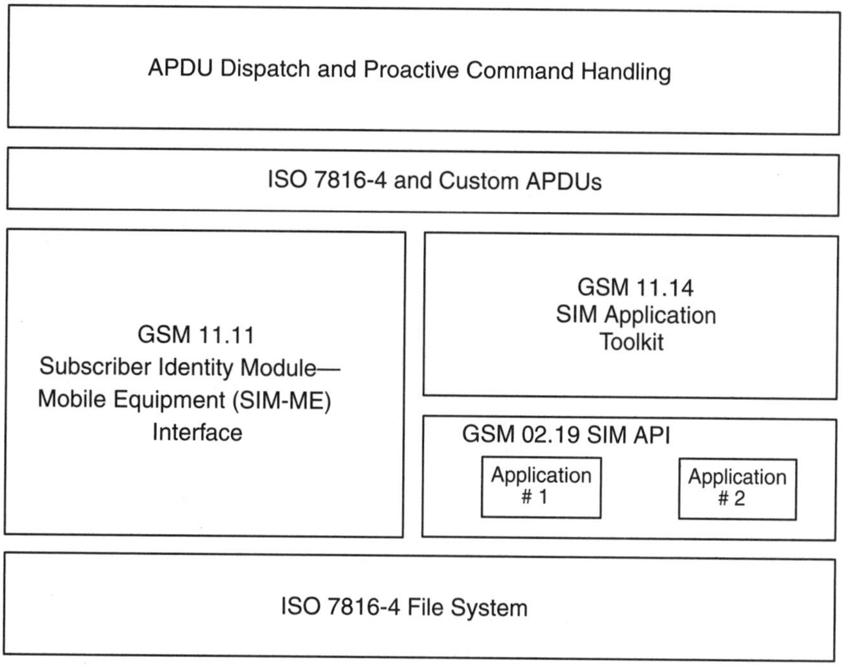

The SIM

Figure 7-4
Architecture of a second generation SIM.

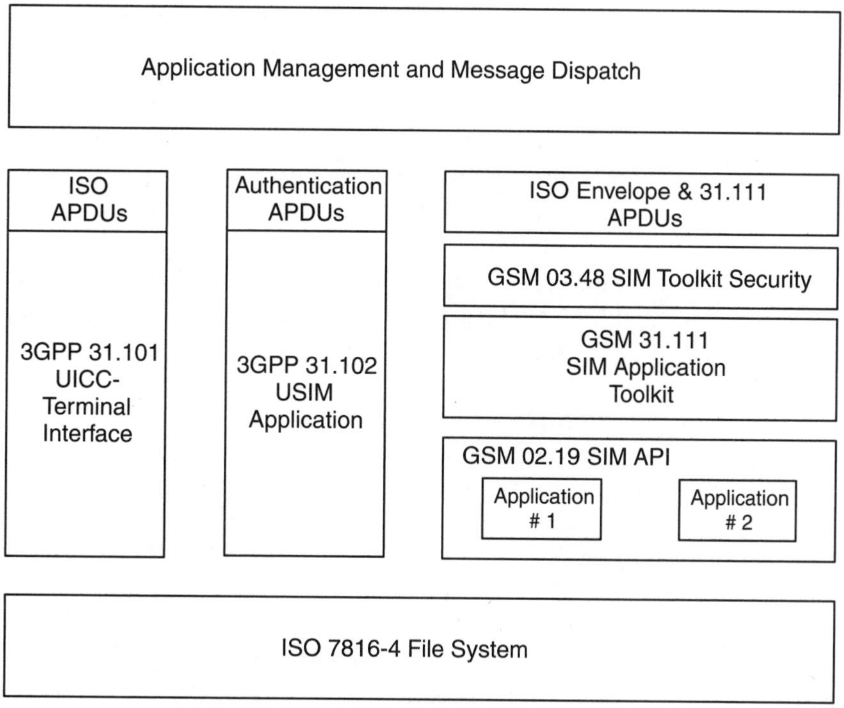

```
┌─────────────────────────────────────────────────────────────┐
│        APDU Dispatch and Proactive Command Handling         │
└─────────────────────────────────────────────────────────────┘

┌─────────────────────────────────────────────────────────────┐
│              ISO 7816-4 and Custom APDUs                    │
└─────────────────────────────────────────────────────────────┘

┌───────────────────────────┐  ┌──────────────────────────────┐
│                           │  │         GSM 11.14            │
│        GSM 11.11          │  │      SIM Application         │
│  Subscriber Identity      │  │         Toolkit              │
│  Module—Mobile            │  ├──────────────────────────────┤
│  Equipment (SIM-ME)       │  │    GSM 02.19 SIM API         │
│       Interface           │  │  ┌───────────┐ ┌───────────┐ │
│                           │  │  │Application│ │Application│ │
│                           │  │  │    #1     │ │    #2     │ │
└───────────────────────────┘  └──┴───────────┴─┴───────────┴─┘

┌─────────────────────────────────────────────────────────────┐
│                 ISO 7816-4 File System                      │
└─────────────────────────────────────────────────────────────┘
```

Figure 7-5
Architecture of a third generation SIM.

```
┌─────────────────────────────────────────────────────────────┐
│         Application Management and Message Dispatch         │
└─────────────────────────────────────────────────────────────┘

┌──────────┐ ┌──────────────┐  ┌──────────────────────────────┐
│   ISO    │ │Authentication│  │     ISO Envelope & 31.111    │
│  APDUs   │ │    APDUs     │  │            APDUs             │
├──────────┤ ├──────────────┤  ├──────────────────────────────┤
│          │ │              │  │ GSM 03.48 SIM Toolkit Security│
│3GPP 31.101│ │ 3GPP 31.102 │  ├──────────────────────────────┤
│  UICC-   │ │    USIM      │  │        GSM 31.111            │
│ Terminal │ │ Application  │  │     SIM Application          │
│Interface │ │              │  │        Toolkit               │
│          │ │              │  ├──────────────────────────────┤
│          │ │              │  │    GSM 02.19 SIM API         │
│          │ │              │  │ ┌───────────┐ ┌───────────┐  │
│          │ │              │  │ │Application│ │Application│  │
│          │ │              │  │ │    #1     │ │    #2     │  │
└──────────┘ └──────────────┘  └─┴───────────┴─┴───────────┴──┘

┌─────────────────────────────────────────────────────────────┐
│                 ISO 7816-4 File System                      │
└─────────────────────────────────────────────────────────────┘
```

There is a fourth generation on the horizon that we will discuss in Chapter 12.

Who Are You?

Throughout its evolution, the SIM has never lost sight of its primary function—to authenticate the subscriber to the network. At its heart, it is an identity token; as a network identity, it has become more, not less, important. Its unique ability to perform this authentication function is what has attracted applications to the SIM platform.

Initially, the SIM could recognize three entities:

- **ADM**—This is the network operator who performs ADMinistration duties on the SIM. ADM might be thought of as the UNIX superuser.
- **PIN**—This is the subscriber, the person toting the phone and the person who enters the PIN to activate the SIM on the phone so that it can get the phone on the network.
- **PIN2**—This typically is the entity paying the bills for use of the mobile. PIN2 is stronger than PIN and can, for example, reset PIN, but it is not as strong as ADM.

Associated with every file on the SIM smart card is a set of rules called an *access control list* that says who can do what to the file. Here is the description for the Fixed Dialing Numbers file, EF_{FDN}, directly from the 3G 31.102 standard (shown on next page).

Good old 31.102 dictates that EF_{FDN} can be read by the person who knows the PIN, it can be updated only by the person who knows PIN2, and it can be deactivated and activated only by individuals who can prove they are ADM.

The FDN table contains a list of the numbers that a phone with this SIM can call. Imagine a phone given by a company to an employee to use on the road to call the office and the supply depot. It's OK for the employee to read the numbers he or she can call (PIN) but obviously only the company itself should be able to add new numbers to the file (PIN2). If PIN had UPDATE rights, then the employee could write whatever numbers they please into the table and this would defeat the purpose of the table.

There is a table like this for every one of the 80 odd files on a SIM card and this is how access to all the data in the SIM is controlled. It's

The SIM

Identifier: 6F3B	Structure: linear fixed	Optional
Record length: X+14 bytes	Update activity: low	

Access conditions:
    READ   PIN
    UPDATE   PIN2
    DEACTIVATE ADM
    ACTIVATE   ADM

Bytes	Description	M/O	Length
1 to X	Alpha identifier	O	X bytes
X+1	Length of BCD number/SSC contents	M	1 byte
X+2	TON and NPI	M	1 byte
X+3 to X+12	Dialling number/SSC string	M	10 bytes
X+13	Capability/configuration2 identifier	M	1 byte
X+14	Extension2 record identifier	M	1 byte

also possible to stipulate the access condition ALWAYS or NEVER. ALWAYS means anybody can do it at any time without proving who they are. NEVER means nobody can do it, no matter who they can prove they are.

Evolution of SIM Standards

The most well-known smart-card SIM is in GSM phones, but the contributions of this portable tamper-resistant token to the security of a mobile telephone network did not go unrecognized by other mobile telephony access technologies. SIMs can also be found in CDMA phones, iDen phones, and TDMA phones. All of these other SIMs took the GSM SIM standards as their starting point and headed off in directions dictated by their particular technologies and business environments.

As a result, SIM standards grew up closely associated with the networks in which they were used and thus the standards development organization (SDO) that defined those networks. Whereas GSM clearly led the way for the use of the SIM chip, other network technologies recognized the value of the SIM in the network architecture and were active at least to some extent in defining the SIM.

Mobile Technology	SDO	SIM Standard	Name of the SIM
TDMA	ETSI	TELEPOINT	TIM
GSM	ETSI	GSM 11.11	SIM
TETRA	ETSI	EN 300 812	TSIM
TDMA—136	TIA/EIA	ANSI-136-510-B	UIM
TDMA—PDC	ARIB	T63-31.102	UIM
TDMA—Rev-C	TIA/TR45.3	SP 4027-030,033, 034	UIM
IMT-2000	ISO	Q.1741	UIM
CDMA	CDG	CDGRF 43	R-UIM
CDMA	3GPP2	IS 820	R-UIM
CDMA 2000	ARIB	T64-C.S0023	R-UIM
W-CDMS	ARIB	31.101	UIM
iDEN	Motorola	Proprietary	SIM
WAP	WAP Forum	WAP-260-WIM-20010712-a	WIM
WAP & SIM	None	None	SWIM
TDMA	3GPP	Proposed	ISIM

In late 1999, a meeting was held in Austin, Texas, to gather support for the notion of a common SIM that could be defined across all telecommunications technologies. This resulted in all the TDMA technologies banding together under the 3GPP flag and a liaison and cross-reference linkage between TDMA and CDMA SIM standardization efforts.

The SIM

Mobile Technology	SDO	SIM Standard
TDMA	3GPP T3	31.102
CDMA	3GPP2	C.S0023
CDMA	TIA/EIA	IS-820

In early 2000 ETSI created a new project called Smart Card Platform, which is currently chartered with defining a common generic SIM for all communication applications.

Mobile Technology	SDO	SIM Standard
UICC	ETSI SCP	102.221

Because all SIM standards take the GSM SIM standard as a starting point and there are more GSM SIMs in the field than all the other SIMs put together, the

```
GSM 11.11 -> 3G 31.101 -> SCP 102.221
```

sequence is the dominant SIM standard track and SCP 102.221 is the one to follow if you follow only one. The agreement in place among the SIM standards groups is that, if a particular access technology such as GSM wants to deviate from SCP 102.221, the standard describing that deviation will be expressed as a delta or modification of SCP 102.221 rather as a cut-and-paste job on 102.221.

Mobile Technology	SDO	Extension of ETSI SCP 102.221
GSM	ETSI	51.111
TDMA	3GPP	31.111

The real revolution that took place in the process of this evolution was that the SIM became a bag of bits rather than a piece of hardware. The common hardware foundation of all SIMs being developed is the UICC. Some people think this stands for Universal Integrated Circuit Card, but it doesn't actually stand for anything. Any number of the existing SIM specifications now run as software applications on this hardware foundation.

In particular, the GSM SIM application running on the UICC hardware base is called the USIM. Because the software USIM and the hardware SIM behave more or less the same from an application developer's point of view, we will continue to refer to the SIM and understand that we are really referring to the USIM application running on the UICC platform as we move into the future.

Another identity application planned for the UICC is the ISIM. The I in ISIM stands for IMS, which in turn stands for IP Multimedia Service (where, of course, IP stands for Internet Protocol). The idea is that the USIM identity application lets you use the mobile network and the ISIM application lets you access multimedia content. The network operator would control the USIM because it's the operator's network. Vivendi, AOL-Time-Warner, or Sony would control the ISIM because it's their content.

The files and directory structure of the UICC and the USIM are shown in Figures 7-6 and 7-7. All of these files are kept in the EEPROM of the UICC and can be accessed using the 22 smart-card commands supported by the UICC operating system.

The Birth of the SIM Application Toolkit

In the early 1990s, as GSM began to take off, operators cast about for ways to attract new subscribers and switch those subscribers from their competitors. Rate plans and special deals proliferated, but there are only so many people you can attract with free calls to Romania on Tuesday. A better idea and certainly a more successful business model was unique supplemental services such as speed dialing and call forwarding. The downside of building some of those services into the network was that scaling killed you. Imagine, for example, the data storage capacity and data processing capability needed to store and provide instant access to millions of personal phonebooks on central servers.

As the operator's engineers looked around for alternative ways to offer these services, their eyes naturally fell on that little computer in every GSM phone. The initial reaction of the security gurus who had designed the SIM was a very emphatic "No applications in the SIM. Full stop."

The SIF

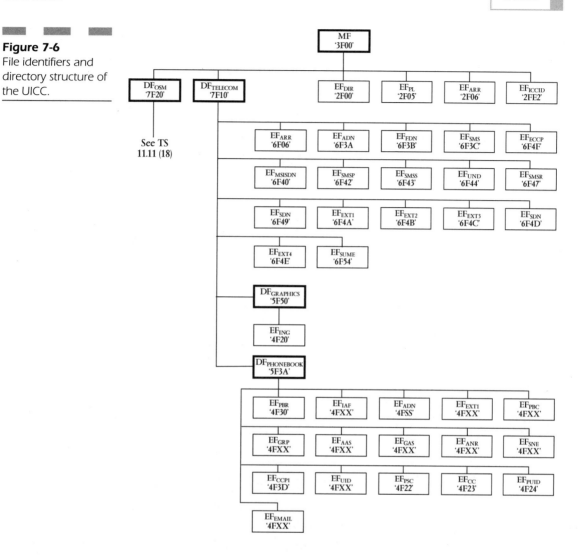

Figure 7-6
File identifiers and directory structure of the UICC.

Figure 7-7
File identifiers and directory structure of the USIM ADF.

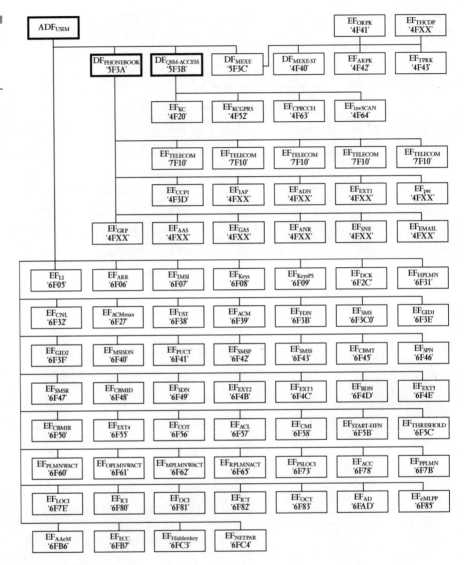

Nevertheless, pressures of business overruled purity of design as usual, and slowly but surely the notion of SIM-based applications came into play. Initially, this involved putting inert data such as frequently called numbers on the SIM, but mold-breaking experiments by Swisscom to use the SIM for real-time call redirection convinced people that the SIM could play a more active role in advanced service delivery. The trouble was the master/slave relationship between the SIM and the handset.

Recall that the communication protocol between the handset and the SIM is for the handset to send the SIM a command and for the SIM to execute the command and give back the response, i.e., a master/slave relationship. Given this relationship, it seemed that every new service offering would entail modification of the handset because the handset had to drive the application and the SIM could only do as it was told. Even if the SIM could give application-specific instructions to the handset, the handset would have to be programmed to understand them. This tied a particular version of the handset to a particular version of the SIM and defeated one of the key features of the SIM, namely portability from one handset to another. Further, the time needed for developing, manufacturing, and delivering new handsets was far too long to respond to rapidly opening and closing market windows for advanced services.

In August 1994 two engineers at BT Cellnet, Colin Hamling and Kristian Woodsend, had a pair of breakthrough ideas. As fate would have it, Kristian worked for Colin at the time, so they could rub their ideas together to light a fire under the evolution of the SIM.

Kristian's idea was to define a standard set of relatively low-level, general-purpose commands on the handset and let SIM assemble those in various ways to create operator-specific services. This was the Right Stuff except something has to be done to turn around the master/slave relationship between the SIM and the handset so that the SIM could tell the handset what to do. Colin's contribution was to define a new smart-card status word that said "The command you gave me executed just fine AND I HAVE A COMMAND FOR YOU." The most elegant and rewarding engineering is doing for a dime what any fool can do for a dollar (Figure 7-8).

Figure 7-8
Proactive command protocol.

The key word was "standardized." Colin and Kristian had to get their proposal written into GSM standards so that every handset supported the commands and the new status word to preserve the portability of the SIM. Colin at the time just happened to be the vice-chairman of the GSM SMG9 standards committee that was in charge of writing such standards so he and Kristen worked furiously through the last week of August to get their idea ready for submission to the SMG9 meeting in early September 1994.

Fortunately for Colin and Kristin, the chairman of SMG9 at the time, Dr. Klaus Vedder, although a security expert and the original champion of the use of the SIM in GSM phones, saw more potential than threat in coupling the SIM into vertical applications. It must be noted in passing that Klaus is still the chairman of the two committees that guide the evolution of the SIM, 3GPP's T3 committee and ETSI's SCP committee, and he is as receptive and encouraging as ever to new ideas.

The SIM Toolkit used the ISO 7816-4 ENVELOPE command to move data from the handset to the SIM. At the same time that the Cellnet notion of a SIM Toolkit, as it was called, was coming into focus in SMG9, Swisscom's notion of SIM data download was winding its way through the SMG9 standardization process. The idea of SIM data download was to use special SMS messages to cause data to be written or updated in the SIM. The leverage provided by linking these two ideas—seamless communication from the network applications to applications on the SIM—was quickly recognized and at the February 1995 SMG9 meeting the Proactive SIM and SIM Data Download work items were merged into the SIM Application Toolkit work item.

The SAT is a happy and cooperative marriage of two innovations: the ability of the SIM to send commands to the handset using a new ISO 7816-4 status word and the ability of the handset to notify the SIM of events using the ISO 7816-4 ENVELOPE command. What must be noted from an engineering design perspective is that this revolution in the use of the SIM was brought about wholly within the existing technical framework and in a way that was completely backward compatible. This, in our opinion, is engineering at its finest.

The SAT API

Very quickly after the SAT had been standardized in the document ETSI TS 11.14, the SIM card manufacturers produced SIMs that enabled the network operators to build SIM Toolkit applications on the SIM. Handsets that supported the SIM Toolkit also began to be available in phone stores on Tottenham Court Road.

As might be expected, when things are moving this quickly, applications developed for one manufacturer's SIM were not easily portable to another manufacturer's SIM. As a result, in late 1996 a new work item was formed within SMG9 to standardize SAT. The rapporteur of this work item was Mark Green from the UK GSM operator, Orange, and the secretary of the work group was Kristian Woodsend who by now had moved to Aspects Software in Edinburgh, Scotland.

In summer 1996, one of the SIM manufacturers, Schlumberger, announced that it had developed a Java virtual machine for smart cards. This virtual machine made writing applications that ran on the smart card much easier than had been the case up to that point. Application programmers could use a familiar programming language and applications could be securely loaded onto, executed, and deleted from the card even after it was in use. Following Schlumberger's lead, almost all of the other smart-card manufacturers implemented Java virtual machines for their smart cards. Thus, when the need arose to standardize an API for SAT, it was natural to use the Java virtual machine as the framework.

The standard ETSI TS 02.19 provides a programming language-independent description of the SAT API and GSM 03.19 binds that generic description to the Java programming language. Initially the Java SIM also had to be an "official" Java Card but that requirement was removed from the 3GPP standard in March 2001. There is also a binding of the C programming language to the SAT API in the standards pipeline.

Just like the SIM itself, there are versions of the SAT that are specific to different access technologies.

Mobile Technology	SDO	Toolkit Standard
GSM	ETSI	GSM 11.14
3G	3GPP	3G 31.111
TDMA Rev-C	TIA TR45.3	Rev-C 136-037
CDMA/CDMA 2000	TIA TR45.4/3GPP2	UATK
TDMA	GAIT	GAIT-H-1-1-2-0

There is an effort underway in the ETSI Smart Card Platform project to unify all these SIM Toolkit standards so that SIM-based applications can move between access technologies without change. The specification is called the Card Application Toolkit, or CAT.

Platform	SDO	Toolkit Standard
UICC	ETSI SCP	ETSI TS 102.223

The USAT Interpreter

Some would argue that application downloading to the SIM is a bridge too far. Certainly some big operators, notably the Vodafone

Group, have stepped back from this way of coupling the GSM phone into IT applications. Although downloaded applications can do a lot, there are three primary drawbacks.

First, downloaded applications occupy precious space in the EEPROM of the SIM. You'd better be sure that they are used frequently enough—and generate enough revenue when they are used—to justify the SIM real estate they demand.

Second, downloaded applications are hard to manage. Not only does the network operator have to be sure that the code is squeaky clean, but the process of getting them to the SIM and activating them once they are there has to be airtight from a security point of view. A downloaded application doesn't have to even touch the edge of the virtual machine's sandbox to set up the occasional call to Siberia Tel and Tel. There is also the central database of keeping track of the different applications on the SIMs of the subscribers.

Third, and perhaps most problematic, is the issue of the human interface. No matter whose application it is, the subscriber can't figure out how to use it will call their network operator. It doesn't take very many of these calls coupled with the expense of training all the call center people on all the applications before the spreadsheet for downloaded applications turns red.

In 1998, Across Wireless (now Sonera SmartTrust) came up with an elegant solution to these problems that still harnessed the full capability of the proactive SIM. They installed a microbrowser on the SIM that could interpret downloaded mark-up language pages. The analog was to World Wide Web browsers but the emphasis was on tickling the SAT API rather than the content of the pages themselves.

The Across Wireless microbrowser addressed the shortcomings of full-bore application download as follows. First, the downloaded pages were thrown away after they were interpreted so that they didn't burn up EEPROM space. Second, mark-up languages were much simpler than procedural languages so they could be checked automatically for deviant behavior. Third, publication of a style manual and quick manual checks when the application was certified could standardize and homogenize the human interface and thereby minimize the number of calls to customer service.

The Across Wireless microbrowser approach got a very favorable reception in the marketplace and in mid-2000 Vodafone proposed that microbrowser technology be standardized. In the meantime the SIM manufacturers, seeing the success of Across, had come up with their own microbrowser based on a Gemplus prototype.

These two approaches were "harmonized" by the standardization process (much like a sausage is a harmonization of its ingredients) and the result is the 3GPP USAT Interpreter.

Standard	Description
3GPP TS 22.112	USAT interpreter—stage 1
3GPP TS 31.112	USAT interpreter architecture description—stage 2
3GPP TS 31.113	USAT interpreter byte codes—stage 3
3GPP TS 31.114	USAT interpreter protocol and administration—stage 3

We explore microbrowsers and the USAT interpreter in depth in Chapter 11.

Summary

In this chapter we introduced you to the GSM SIM. It's a small computer but very well connected. The SIM began life as a smart card that held the secret key to identifying the subscriber to the network and connecting that individual to a billing account. Thus the name Subscriber Identity Module.

As the demand for more comprehensive mobile applications grew, the SIM morphed into an full-fledged application platform. While staying within the technical framework of the smart card, SIM engineers have used this framework creatively to add capabilities such as an internal API, the ability to support multiple high-level languages, event notification, and proactive interaction with the handset.

In the following chapters, we consider in detail how to build applications for the SIM and how to communicate with them using SMS.

CHAPTER 8

SIM Toolkit API: Proactive Commands and Event Download

All the application development we've discussed so far has treated the SIM as a black box. We sent it messages and we responded to what it sent back, but we were definitely on the outside looking in. Now it's time to climb inside the SIM and be on the inside looking out.

In this chapter we discuss the fundamentals of building applications on the SIM itself. In fact, code on the SIM is probably only part of a distributed application that might work with codes on the handset, network servers, and certainly Internet servers. We will describe the environment in which the code on the SIM works and leave it as an exercise for you the reader to get it working with the other parts of the application.

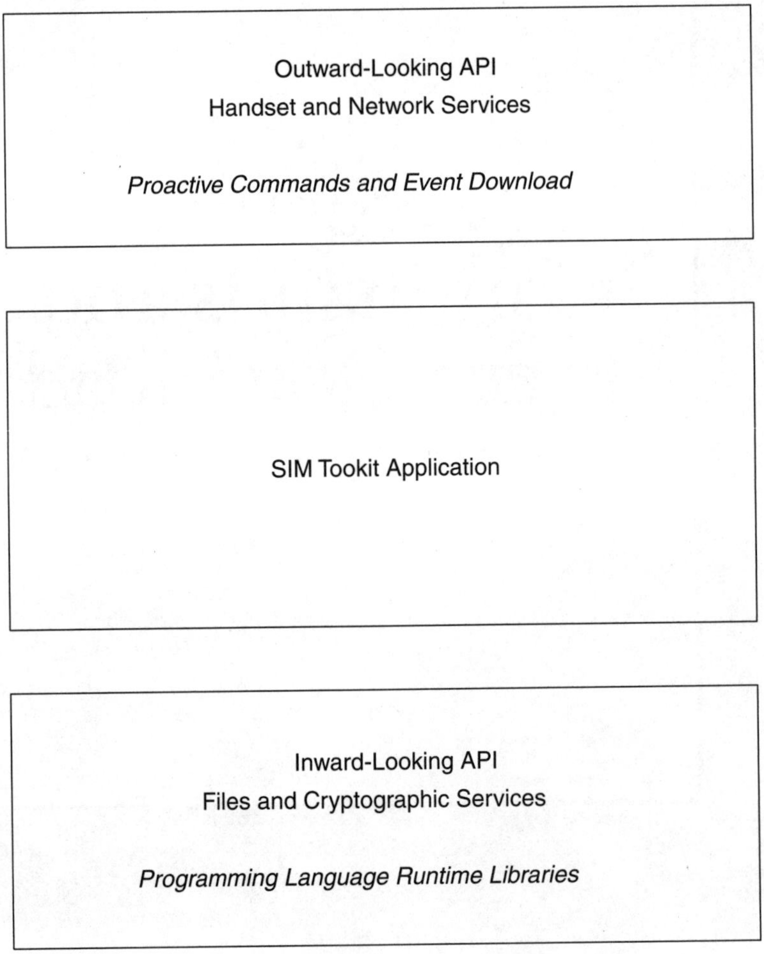

Figure 8-1
Inward and outward APIs for SIM applications.

Outward-Looking API
Handset and Network Services

Proactive Commands and Event Download

SIM Tookit Application

Inward-Looking API
Files and Cryptographic Services

Programming Language Runtime Libraries

As shown in Figure 8-1, a SIM application has to deal with two APIs. The inward-looking one provides standard, albeit small, operating system services to the application such as file reading and writing and computational functions such as cryptographic calculations. The outward-looking one connects the SIM application to the human interface capabilities of the handset and to the network.

The inward-looking API is typically language-specific and offers scaled-down versions of familiar functions. We'll assume that you know how to open, read, write, and close files in your favorite programming language, so we won't spend much time on this API.

It is the outward-looking API that makes a SIM application different from an application you might write for a laptop, a PDA, or even a handset. Because a SIM application is running in a secure, controlled, and trusted environment owned by the network operator, it can do more than an application running in an insecure, uncontrolled, and untrusted environment such as the handset. Further, this outward-looking API has been carefully standardized and thus provides portability to applications.

There are two types of information flow between your SIM application and the outside world. The only difference is who initiates the conversation. If the SIM initiates the conversation, then the flow is called a "proactive command"; your application is asking the handset to do something. If the handset initiates the conversation, then the flow is called an "event download"; the handset is the application that made something happen (Figure 8-2).

The two flows together comprise the SIM Toolkit API and this API is the interface between your SIM application and the outside world. It is also the API that we are going to discuss in this chapter.

Proactive Commands

A proactive command, as we discussed briefly in the previous chapter, is a command from the SIM application to the handset asking it to do something on your behalf. It is called proactive because, uncharacteristically, the SIM is initiating the communication.

As of late 2001, there are 31 proactive commands on the SAT API. The SIM Toolkit proactive commands are listed in Table 8-1. The most recent list and the complete documentation are always available in ETSI TS 102.223.

Figure 8-2
Reactive interface between handset and SIM.

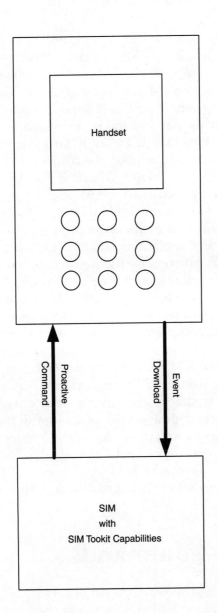

TABLE 8-1

SIM Toolkit Proactive Commands

SIM Toolkit Proactive Commands
DISPLAY TEXT
GET INKEY
GET INPUT
MORE TIME
PLAY TONE
POLL INTERVAL
REFRESH
SET UP MENU
SELECT ITEM
SEND SHORT MESSAGE
SEND SS
SEND USSD
SET UP CALL
POLLING OFF
PROVIDE LOCAL INFORMATION
SET UP EVENT LIST
PERFORM CARD APDU
POWER OFF CARD
POWER ON READER
GET READER STATUS
TIMER MANAGEMENT
SET UP IDLE MODE TEXT
RUN AT COMMAND
SEND DTMF
LANGUAGE NOTIFICATION
LAUNCH BROWSER
OPEN CHANNEL
CLOSE CHANNEL
RECEIVE DATA
SEND DATA
GET CHANNEL STATUS

When the SIM has a command to send to the handset, it must wait until it is asked to do something pedestrian like SELECT FILE or READ BINARY. When this task is complete, rather than responding with a status word of 0x9000, the SIM responds with 0x91xx. As mentioned in Chapter 6, this status word means "The command you gave me executed just fine and I have a command of xx bytes for you." In its own time, the handset sends the SIM a command explicitly designed for the proactive SIM, FETCH. FETCH is a very simple APDU:

CLA	INS	P1	P2	Lc
0x80	0x12	0x00	0x00	Number of bytes in command (the *xx* in 0x91xx)

The FETCH command pulls the SIM's command onto the handset, where it is parsed and executed (Figure 8-3).

The handset then tells the SIM how things went. It does this with another specially designed command, TERMINAL RESPONSE. In this command, the handset passes back to the SIM the results of executing the SIM's command.

CLA	INS	P1	P2	Lc	Data
0x80	0x14	0x00	0x00	Number of bytes in the following data field	Result of executing the proactive command

The data returned from the handset to the SIM obviously depends completely on what the SIM asked the handset to do. We will cover what comes back as we discuss each proactive command.

As you probably noticed by looking over the SIM commands in Table 8-1, one of the commands that the SIM can give to the handset, POLL INTERVAL, is a command that asks the handset to poll the SIM regularly. The handset uses the APDU STATUS to perform this polling function when it doesn't have anything else for the SIM to do. This lets the SIM give the handset commands on a regular schedule rather than just when the handset has work for it.

The world would be a wonderful place if the SIM gave the handset APDUs just like the handset gave the SIM.

SIM Toolkit API: Proactive Commands and Event Download

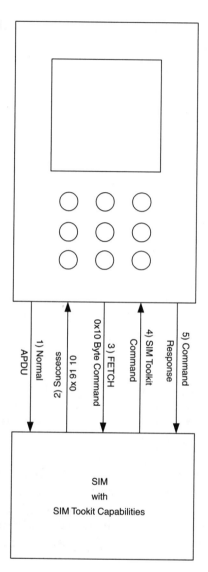

Figure 8-3
Proactive command protocol.

Unfortunately, it didn't turn out that way. What the SIM gives the handset is sequences Tag-Length-Values (TLVs) and it's a sequence of bytes that look like this:

Tag	Length	Value
Kind of data	How many bytes follow	Data bytes

In the most general definition (ASN.1/ISO/IEC 8825), the tag and the value can be any number of bytes, with the details of their encoding telling you exactly what is current. In the world of the SIM, the tag is always 1 byte and the length is 1 or 2 bytes, encoded as follows:

Length	Length Value	Byte 1	Byte 2
1 byte	0 to 127	Length— 0x00 to 0x7F	
2 byte	128 to 255	0x81	Length— 0x80 to 0xFF

(You certainly can see how quickly you can get your pants in a bunch in these matters.)

The Value can itself be yet another TLV. If it isn't, if it is just data then the TLV is called simple TLV. If the data part in turn is a TLV (*ad nauseum*), then the TLV is called a compound TLV.

A TLV obviously is a very general and very flexible way to represent data, but its utility turns on the tags having well-defined and unambiguous meanings. In the world of smart cards in general and SIMs in particular, the tags are very carefully regulated. ISO 7816-6 is a long list of recognized tag values and their definitions.

The meaning of a tag often depends on context and many are with the SIMs. Even when the tag is context dependent, it is well-defined and unambiguous within the recognized context. Section 9.3 in ETSI TS 102.223 is a long list of all the tags used in the traffic back and forth between the SIM and the handset. Even though the TLV representation is not a compelling format for human understanding, it is a very compelling format for machine processing, which after all is the name of the game.

To get our TLV feet wet, let's take a look at one of the commands that the SIM can send to the handset. It is the canonical command example, PLAY TONE. Now you would think that this was an instruction to play say middle C or G sharp above middle C but nooooooooooo. It's all about telephony tones like dial tones and call waiting.

The PLAY TONE command is a compound TLV that looks like the data shown in Table 8-2, its grisly, byte-by-byte details.

SIM Toolkit API: Proactive Commands and Event Download

TABLE 8-2

PLAY TONE Proactive Command Example

Byte Number	Byte Value	Comments
1	0xD0	Tag—proactive SIM command TLV
2	0x15	Length—the value following the proactive command TLV is 21 bytes long
3	0x01	Tag—command details TLV
4	0x03	Length—the value following the command details TLV is 3 bytes long
5	0x01	Value—command number—Integer identifier of this command to the handset
6	0x20	Value—type of command—PLAY TONE command
7	0x00	Value—command qualifier—no qualifier for this command
8	0x02	Tag—device identity TLV
9	0x02	Length—the value following the device identity TLV is 2 bytes long
10	0x81	Value—the source device of the command is the UICC
11	0x82	Value—the destination device of the command is the handset
12	0x05	Tag—alpha identifier TLV
13	0x03	Length—the value following the alpha identifier TLV is 3 bytes long
14	0x42	Value—"B"
15	0x4F	Value—"O"—Display "BOO" when you play the tone
16	0x4F	Value—"O"
17	0x0E	Tag—tone TLV
18	0x01	Length—the value following the tone TLV is 1 byte long
19	0x01	Value—the coding for the dial tone tone, i.e., play the dial tone

continued on next page

TABLE 8-2

PLAY TONE Proactive Command Example (continued)

Byte Number	Byte Value	Comments
20	0x04	Tag—duration TLV
21	0x02	Length—the value following the duration TLV is 2 bytes long
22	0x01	Value—the unit of the duration is seconds
23	0x05	Value—play the tone for five of the above time units

You'll have to admit this is pretty efficient coding. In actual fact, the alpha identifier, tone, and duration TLVs are optional but there is no problem tacking them on because their tags merely announce their presence.

After the handset blasts the dial tone into the subscriber's ear for 5 seconds (with who knows what result), the handset sends a status response back to the SIM telling it how things went with its command. This is done with a TERMINAL RESPONSE APDU. In our case, the APDU looks like this:

CLA	INS	P1	P2	Lc	Data
0x80	0x14	0x00	0x00	0x0C	0x01 0x03 0x0x 0x20 0x01 0x02 0x02 0x82 0x81 0x03 0x01 0x00

The data field is a TLV that doesn't mean, "He's wondering what the heck is going on." The data field consists of three TLVs, as shown in Table 8-3.

SIM Toolkit API: Proactive Commands and Event Download

TABLE 8-3
Terminal Response

Byte Number	Byte Value	Comments
3	0x01	Tag—command details TLV
4	0x03	Length—the value following the command details TLV is 3 bytes long
5	0x01	Value—command number—integer identifier of this command to the handset
6	0x20	Value—type of command—PLAY TONE command
7	0x00	Value—command qualifier—no qualifier for this command
8	0x02	Tag—device identity TLV
9	0x02	Length—the value following the device identity TLV is 2 bytes long
10	0x82	Value—the source device of the command is the handset
11	0x81	Value—the destination device of the command is the UICC
12	0x03	Tag—result TLV
13	0x01	Length—the value following the Result TLV is 1 byte long
14	0x00	Value—command was performed successfully

The command details TLV duplicates exactly the command details TLV in the proactive command to which the result pertains. The device identities TLV reverses the direction of the flow between the devices. The result TLV says what happened. In this case, the value says everything went fine. In other words, this compound terminal response TLV handset talk for 0x9000.

Just to drive home the point, Table 8-4 shows what the entire conversation between the SIM and the handset would look like:

TABLE 8-4

Summary of Proactive Command Traffic

Direction	Bytes	Meaning
Handset to SIM	0x80 0xF2 0x00 0x00	STATUS command—"What's happening'?"
SIM to handset	0x9117	Status word—"I've got 23 bytes for you."
Handset to SIM	0x80 0x12 0x00 0x00 0x17	FETCH Command—"Sock 'em to me"
SIM to handset	0xD0 0x15 0x01 0x03 0x01 0x20 0x01 0x02 0x02 0x81 0x82 0x05 0x03 0x42 0x4F 0x4F 0x0E 0x01 0x01 0x4 0x02 0x01 0x05 0x90 0x00	"Here they are!"—Notice the 0x9000 OK at the end; this is a response to an APDU, after all
Handset to SIM	0x80 0x14 0x00 0x00 0x0C 0x01 0x03 0x0x 0x20 0x01 0x02 0x02 0x82 0x81 0x03 0x01 0x00	TERMINAL RESPONSE command—"Command performed successfully"
SIM to handset	0x9000	Status Word—"OK, thanks. Nothing else for you to do right now."

Details of SIM Toolkit Commands

There are literally hundreds of options and flags among the 31 proactive commands, all carefully coded as TLVs. Clearly we can't cover them all. Good old 102.223 is your best reference ... actually it's your only reference. What we can do is give you a little sketch of each of the proactive commands. We'll divide the SIM Toolkit proactive commands into four categories:

1. **Application commands**—used to build typical SIM Toolkit applications
2. **Smart-card commands**—used to interact with another smart card plugged into the handset
3. **General communication commands**—a general-purpose interface to all the various bearers that the handset supports

SIM Toolkit API: Proactive Commands and Event Download

4. **System commands**—used to stay synchronized with the handset and the network

Application Commands

DISPLAY TEXT. Not surprisingly, this command produces a scrap of text on the screen along with an icon, if you like. You can also decide how long the text should be displayed. The text string can be up to 240 bytes and displayed in a number of different alphabets. The TERMINAL RESPONSE includes an indication of how the subscriber reacted to the displayed text, not emotionally but whether they hit one of the function keys or did nothing at all.

GET INKEY. This command produces some text and an optional icon and requests a single key hit from the subscriber, which it returns in the TERMINAL RESPONSE. You can respond with yes or no, a digit, or any of the various alphabetic characters available.

GET INPUT. This command is like GET INKEY except that you can get back a bunch of characters. You can set the minimum and maximum number of characters you want. TERMINAL RESPONSE includes the number of characters and the characters themselves.

SETUP MENU. There are many menus on the handset, but one is reserved as a starting point for applications on the SIM. This is called the setup menu and this proactive command sets up the setup menu (sorry, it was to obvious to pass up). It's up to the handset to work this SIM starting point menu into all the other menus it handles for its own features.

Note that this menu doesn't become a permanent part of the menu system on the handset. It goes away when the handset is turned off, so this command has to be sent each time the SIM is powered up.

The menus included in the handset by the handset manufacturer or the network operator are completely different from the menus generated by the SIM. Because the SIM wants to move easily between handsets, it has to carry its menu structure with it. The SETUP MENU command produces an entry in a handset set menu, typically the top-level menu and, when this entry is picked, the SIM Toolkit menu contained in that command is displayed.

SELECT ITEM. This is the workhorse proactive command. It calls up a menu of items and you pick one. You can supply a title and an icon to identify the menu.

SEND SHORT MESSAGE. Now we're talking power application. One of the advantages of SIM applications over handset applications is that you can get onto the network. With this proactive command, the application can send an SMS_SUBMIT with text or binary data. It's perfect for sending an encrypted transaction back to action central.

SETUP CALL. Yes, you can open up a voice channel from your SIM application. You may want to do this as part of some convenience or productivity application and immediately turn the call over to the subscriber. Or you may want to call a machine of some sort and use the SEND DTMF command to talk to it in touch tones. Or you could combine the two and have the SIM application hack its way through the "touch 1 if you were born in an odd numbered month" jungle and then turn the call over to the subscriber when you find a human being.

SEND DTMF. Sends a touch tone on the voice channel opened up by SETUP CALL.

PLAY TONE. We covered this above; basically, it lets you play a telephony tone such as a dial tone or a busy signal for the subscriber. Pity it's not an API to the iMelody capability of the handset we discussed in the SMS chapters, eh?

PROVIDE LOCAL INFORMATION. It turns out that the handset knows a lot about what's happening out there. PROVIDE LOCAL INFORMATION is the way that your application asks the handset what's going on. The handset can tell your application about the network it's on, how strong the signals it is seeing are, what cells it is talking to, and other useful information such as the current time and date and time zone.

You can use this information to perform location-based services, but how to do this and how accurate you can get is a topic for another book. Check out www.simtrack.com for an example of how it can be used.

SET UP EVENT LIST. Not only does the handset know a lot, a lot is happening in there. Calls are being made, SMSs are being sent, keys are being hit, screens are being refreshed, Bluetooth and IrDA messages are arriving, and, on some handsets, smart cards are being inserted into second card slots. The SET UP EVENT LIST command is a way for the SIM to tell the handset what events that it, the SIM, would like to be told about. The handset does this with EVENT DOWNLOAD APDUs, which we discuss below.

SET UP IDLE MODE TEXT.   Like SETUP MENU, where you integrate a menu into the handset's set of stock menus, this command lets you add a text string to the stuff that the handset puts on the screen when nothing is happening. Some handsets don't support this command and some network operators don't allow it.

RUN AT COMMAND.   Remember in Chapter 2, where we sent AT commands to the handset to make things happen? Well, you can send these same AT commands from the SIM rather than from a PC connected to the handset and cause the same things to happen as if the command were sent from terminal. The TERMINAL RESPONSE returns the result of running the AT command.

This is really a very elegant way of opening up the capabilities to your SIM application without building sets of new proactive commands. If you take a look at 3GPP TS 27.007, you'll find that you can set a lot of the parameters of the handset, retrieve information over and above that provided by PROVIDE LOCAL INFORMATION, and access the phone book on ... the SIM.

It may seem a little strange to be using the handset to talk to yourself but negotiating the phone book can be a daunting task so you might as well harness all the code on the handset that knows about the phone book structure rather than trying to replicate it in your application. You can also use this technique to get at SMS messages stored on the SIM and the handset.

TIMER MANAGEMENT.   Smart cards including the SIM have no notion of time, i.e., they have no internal timers. Some applications and some communication protocols require a sense of time. You can imagine having to respond to a pop-up question within 5 seconds or timing out a message sent on the Internet. Although it has no built-in timers, the SIM can use the timers on the handset. The TIMER MANAGEMENT command is the reason that a SIM application can start a timer on the handset. The TERMINAL RESPONSE just says that the start has been successful. A later EVENT DOWNLOAD (see below) signals the expiration of the set time.

Smart-Card Proactive Commands

The hope of SIM manufacturers is that there will be an outside slot on the handset that will accept a normal-sized smart card. The idea is that you would use this slot to insert you American Express Blue card or your national identity smart card in order to engage in mobile commerce or to add additional security to your authentication.

A question that comes to mind is, What happens if that card also is running an application? Won't the second card and the SIM fight for the attention of the subscriber? And how does the subscriber know which they are interacting with? Good questions. The realization that the mobile handset is in fact an entire mobile network that supports applications running on all of its nodes is only now dawning on folks. These questions are being hotly debated in a number of standards forums.

POWER ON CARD AND POWER OFF CARD. These two commands toggle power to the second (or third or fourth) smart card.

GET READER STATUS. This returns the status of one of the other card readers attached to the mobile.

PERFORM CARD APDU. This sends a complete APDU to another smart card. The TERMINAL RESPONSE, is not surprisingly, the response of the other smart card. Can you send an APDU to yourself? Check it out.

LAUNCH BROWSER. You can pass a URL to the browser on the handset but this will end your application and pass control of the session to the handset browser. The handset browser will resolve the URL and take it from there.

General Purpose Communication Commands

As the number of modes of communication that the handset can access grew, it became clear that it would be unwieldy if each one created its own set of proactive commands. As a result, a general set of proactive communication commands was designed that would be a uniform interface to all bearers, local and network, slow speed and

SIM Toolkit API: Proactive Commands and Event Download

high speed, circuit and packet switched. These are the proactive commands that are included in the general purpose communication commands category.

OPEN CHANNEL. This command opens a channel on a particular bearer. The command takes a long list of parameters that describe exactly what flavor and color of channel you'd like. The TERMINAL RESPONSE gives you information about the established channel if the handset was successful in opening it and the reason for failure if it wasn't.

SEND DATA AND RECEIVE DATA. The commands used to send data on an open channel and receive data from an open channel.

CLOSE CHANNEL. Closes an open channel.

GET CHANNEL STATUS. Returns TERMINAL RESPONSE information about the current state of an open channel.

System Commands

MORE TIME. Your application shouldn't run more than a couple of seconds without giving the handset a chance to use the SIM for its primary task, which is keeping the mobile on the network. At 3.5 MHz on an Intel 8051, that's roughly 1 million instructions so it's not like you don't have some computer resource to work with. But if you think you're getting near the edge, you can issue a MORE TIME command to give the handset a slice and then pick up where you left off when the TERMINAL RESPONSE to the MORE TIME comes back.

POLL INTERVAL AND POLLING OFF. These toggle the polling of the SIM by the handset and set the frequency with which the handset sends STATUS commands to the SIM when polling is on.

SEND SS AND SEND USSD. SS is Supplemental Service and USSD is Unstructured Supplemental Service. These messages go directly to the network operator's *home location register* (HLR) and access operator-provided services and subscriber features. Unless you are building an application for a network operator, it is unlikely that you will be tapping into these services.

LANGUAGE NOTIFICATION. Just in case you want to switch to, say, Finnish on the fly.

REFRESH. One of the things an SAT application can do is update the files on the SIM; as you recall, this was one of the Ur-applications of the toolkit. For efficiency purposes, when you turn on the handset, it acquires a lot of information from the SIM so it can get at it quickly. If your application changes some information on the SIM, it has to give the handset a nudge to tell it to refresh its caches. The REFRESH command is the nudge.

Event Download

Besides giving the handset commands, the SIM can register for events that it wants to be told about. It uses the SET UP EVENT LIST proactive command to do this (Figure 8-4).

Table 8-5 shows the 19 currently available events that the handset communicates to the SIM.

TABLE 8-5

Event Downloads

Name	Description
SMS-PP	A pp SMS message has arrived
CELL BROADCAST	A cell-broadcast SMS message has arrived
MENU SELECTION	A selection has been made from the main SIM menu
CALL CONTROL	Subscriber is placing a voice call
SMS CONTROL	Subscriber is sending a short message
TIMER EXPIRATION	One of the timers you set has gone off
MT CALL	A voice call is coming in
CALL CONNECTED	A voice call is connecting
CALL DISCONNECTED	A voice call is disconnecting
LOCATION STATUS	We've changed locations
USER ACTIVITY	She's punching the keys again

continued on next page

SIM Toolkit API: Proactive Commands and Event Download

TABLE 8-5
Event Downloads (continued)

Name	Description
IDLE SCREEN AVAILABLE	There's nothing happening on the screen
CARD READER STATUS	Something is happening over on the external smart-card reader
LANGUAGE SELECTION	She's decided to talk in tongues
BROWSER TERMINATION	The handset's browser has been shut down
DATA AVAILABLE	There's incoming data on one of your data channels
CHANNEL STATUS	One of your data channels has changed states
ACCESS TECHNOLOGY CHANGE	We've roamed into a new kind of network
DISPLAY PARAMETERS CHANGED	She's resized the screen

In all cases, the handset uses the ISO 7816-4 ENVELOPE APDU to send a description of the event to the SIM. The ENVELOPE APDU was part of the original set of ISO interindustry commands engraved in stone by ISO 7816-4. It is the "escape hatch" that you find in all good designs. ENVELOPE is a way to get an arbitrary blob of data into the card without telling the card what to do with it, as you do when using UPDATE BINARY, for example.

Here's what the ENVELOPE APDU looks like:

CLA	INS	P1	P2	Lc	Data
0x80	0xC2	0x00	0x00	Number of bytes in following data field	Event data

You can see that there is nothing to it. Basically it says to the card "Here's a bunch of data, you figure out what to do with it." This means, of course, that all the action is in the data blob.

Figure 8-4
Sources of Download Events.

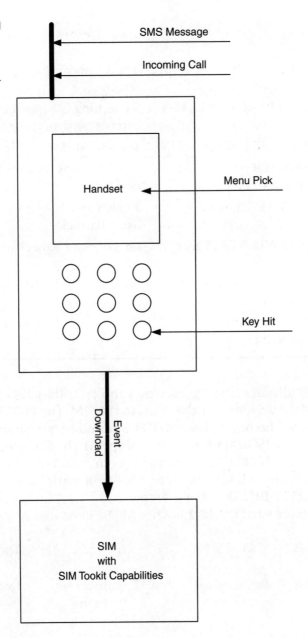

SIM Toolkit API: Proactive Commands and Event Download

For the event downloads, the data blob is a compound TLV wherein the first byte is a tag that says which of the events (for which the SIM has registered interest using the SET EVENT LIST proactive command) has happened and therefore which of the events is described by the simple TLVs that comprise the value portion of the compound TLV.

One of the most useful event downloads is SMS-PP. It is a way of communicating directly with the SIM using SMS. Table 8-6 shows what an EVENT DOWNLOAD for an incoming SMS-PP event might look like.

TABLE 8-6

SMS_PP Event Download

Byte Number	Byte Value	Comments
1	0xD1	Tag—SMS-PP event download
2	0x2A	Length—the value following the SMS-PP event download TLV is 42 bytes long
3	0x02	Tag—device identity TLV
4	0x02	Length—the value following the device identity TLV is 2 bytes long
5	0x82	Value—the source device of the command is the handset
6	0x81	Value—the destination device of the command is the UICC
10	0x03	Tag—address
11	0x07	Length—the value following the address TLV is 7 bytes long
12	0x091	Value—TON/NPI—international ISDN number
13—18	0x91 0x71 0x09 0x57 0x10 0xF0	Value—1 917 907 5010—SMS is from the Omnipoint/VoiceStream/Deutsche Telekom SMSC
19	0x0B	Tag—SMS TPDU
20	0x18	Length—the value following the SMS TPDU (SMS_DELIVER) TLV is 24 bytes long

continued on next page

TABLE 8-6

SMS_PP Event Download (continued)

Byte Number	Byte Value	Comments
21—44	0x04 0x0B 0x91 0x71 0x18 0x53 0x04 0x00 0xF9 0x00 0x00 0x10 0x30 0x11 0x10 0x64 0x54 0x58 0x05 0xC8 0x32 0x9B 0xFD 0x06	Value—SMS_DELIVER

Because an event download is an APDU, the handset should receive a reply and here's a particularly slick part of the SMS-PP event download. The handset automatically sends the reply of the SIM back to the originator of the SMS in an RP_ACK or RP_ERROR message depending on whether the status word from the SIM is 0x9000. By adding some data to the reply, we achieve a two-way link between the SIM and a server on the network. We will see that this is very, very handy when we start to build applications on top of the SAT.

The MENU SELECTION event download signals a pick from the main menu and is a way of starting at the top of an application.

The control event downloads, CALL CONTROL and SMS CONTROL, let the SIM take a look at outgoing activity from the handset and perhaps block it or reroute it. A corporate SIM, for example, might only allow SMSs to go back to the corporate server and only allow long-distance calls to go to corporate locations.

The TIMER EXPIRATION event download tells you that one of the timers you set using the TIMER MANAGEMENT proactive command has gone off (Table 8-7).

SIM Toolkit API: Proactive Commands and Event Download

TABLE 8-7

TIMER EXPIRATION Event Download

Byte Number	Byte Value	Comments
1	0xD7	Tag—TIMER EXPIRATION event download
2	0x0F	Length—the value following the TIMER EXPIRATION event download TLV is 15 bytes long
3	0x02	Tag—device identity TLV
4	0x02	Length—the value following the device identity TLV is 2 bytes long
5	0x82	Value—the source device of the command is the handset
6	0x81	Value—the destination device of the command is the UICC
7	0x24	Tag—timer identifier
8	0x01	Length—the value following the timer identifier is 1 byte long
9	0x01	Value—timer 1 went off
10	0x25	Tag—timer value
11	0x03	Length—the value following the timer value is 3 bytes long
12	0x08	Hour
13	0x00	Minute
14	0x00	Second

The DATA AVAILABLE event download tells you that data has arrived on one of the open channels. It is one of a new breed of event downloads because, rather than having a tag that denotes which event is being downloaded, it uses the generic event download tag, and then uses a simple TLV to tell you which event is being downloaded. Table 8-8 shows what the DATA AVAILABLE event download looks like.

TABLE 8-8

DATA AVAILABLE Event Download

Byte Number	Byte Value	Comments
1	0xD6	Tag—event download
2	0x0F	Length—the value following the TIMER EXPIRATION event download TLV is 15 bytes long
3	0x19	Tag—event List
4	0x01	Length—the value following the event list TLV is 1 byte long
5	0x09	Value—a data available event has occurred
6	0x02	Tag—device identity TLV
7	0x02	Length—the value following the device identity TLV is 2 bytes long
8	0x82	Value—the source device of the command is the handset
9	0x81	Value—the destination device of the command is the UICC
10	0x38	Tag—channel status
11	0x02	Length—the value following the channel status is 2 bytes long
12	0x02	Value—data is on channel 2
13	0x00	Value—no further information
14	0x25	Tag—data length
15	0x01	Length—the value following the data length value is 3 bytes long
16	0x08	Value—8 bytes have come in on this channel

The event list TLV could have contained more than one event but only one of each type. If there were more than one event in the event list, then there would be TLVs following to describe each one. This is a nice efficiency evolution of event download because now one download can carry information about more than one event.

Summary

In this chapter we covered the nitty-gritty details of the interaction between the handset and the SIM. We've seen how the SIM can give commands to the handset and how the handset can tell the SIM about various events that take place. From the point of view of the SIM, the handset is just a docking station. It contains the human interface, access to the network, and connections to other devices and local bearers. To be sure the SIM couldn't do much without the handset in the way your head couldn't do much without your body. But the SIM knows who's the brains and who's the brawn.

In the following chapters, we take a higher-level view of this interactive interface between the SIM and the handset. You can come back to this chapter and dig into ETSI TS 102.223 if you want to remind yourself of how it really works under the hood.

Before we do that, we'll take a closer look at the SMS messages carrying data to the SIM.

CHAPTER 9

End-to-End Security for SMS Messages

In Chapter 8 we discussed how one could set up a communication connection between an application server and the SIM by using SMS and SMS-PP data download. You will recall that the response to this SMS by the SIM went all the way back to the server. This is obviously a very powerful capability except for one small item—security.

The application server is in some physically secure computer center somewhere. The SIM is a tamper-resistant, PIN-protected application platform connected to the wireless network and a modest human interface. The air link is secured with session key encryption. So why worry?

When it was just the network operator sending SMSs to the SIM, there was no need to worry. The operator owned the Over-the-Air (OTA) entity sending SMS messages to the SIM and everything in between. But when entities beyond the network operator wanted to send SMSs to the SIM, in particular to applications running on the SIM SAT, they were concerned.

What they didn't have was end-to-end security. The two endpoints, the application server and the SIM, don't share a security context. If there is a landline link between the server and the SMSC, then it isn't secure. And even if there isn't, the GSM network in general knows the air link session key, so the SMS content is essentially in the clear for anybody who might gain access to this key.

Fortunately, this problem has been thoroughly thought through and the infrastructure to implement its solution enshrined in ETSI ETSI TS 03.48, Security Mechanisms for the SIM stage 2. Although it was designed for use with SAT, 03.48 can be used to provide end-to-end security for any SMS message going to or coming from the SIM.

Remember, in Chapter 3, where we talked about the optional feature descriptions that follow the mandatory header on an SMS message but precede the actual payload of the message? One of the optional feature descriptions that we mentioned but did not elaborate was SIM Toolkit Security. We promised a complete discussion in Chapter 9 and ... well, here we are.

The optional feature description TLV describing SIM Toolkit Security is really just a head fake because all it really does is say that the beginning of the payload contains a description, called a Command Header, which says how the rest of the payload is secured. This trip of two steps is necessary because the entire payload, the Command Header plus the secured data (together called the Command Packet), has to be specified completely and composed at the endpoints, which are the SIM and the application server. The optional feature section of

End-to-End Security for SMS Messages

an SMS message typically is composed by the SMSC or the handset, and users of the SMS channel have different degrees of control over the details of the encoding of optional features.

The SIM Toolkit Security optional feature description TLV, which signals that the user data after all the optional feature descriptions, consists of a ETSI TS 03.48 Command Packet that is simplicity itself:

Byte Number	Byte Value	Comments
1	0x70	Tag—command packet TLV
2	0x00	Length—no value field associated with this TLV

This TLV appears in the User Data Header section of the SMS_SUBMIT prepared by the application server or the SIM and the SMS_DELIVER that arrives at the other end.

All of which (at long last) brings us to the Command Header itself and the following data whose security it describes. The first two bytes of the Command Header denote the total length of the Command Packet payload, i.e., the Command Header plus the secured data. The next byte is the length of the rest of just the Command Header. Figure 9-1 shows how you would construct an SMS message with 03.48 SIM Toolkit Security.

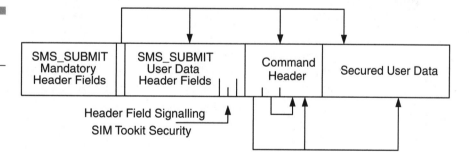

Figure 9-1
SMS TPDU with SAT security.

The image to have in mind here is that the secured data part of the payload is encrypted and the Command Header says how it is encrypted. There are other possibilities and we will discuss them all later but this is the general idea.

The remaining part of the Command Header consists of seven fields:

Field Name	Field Symbol	Length in Bytes	Description
Security Parameter Indicator	SPI	2	A collection of flags describing the type of security imposed on the secured data
Ciphering Key Identifier	KIc	1	Reference to the key and algorithm used to encrypt the secured data
Key Identifier	KID	1	Reference to the key and algorithm used to compute the redundancy check (RC), cryptographic checksum (CC), or digital signature (DS) of the secured data
Toolkit Application Reference	TAR	3	Nominally like an Internet port in indicating which application should handle the secured data but has become overloaded with miscellaneous other uses
Counter	CNTR	5	Replay detection and sequence integrity counter
Padding Counter	PCNTR	1	The number of padding bytes at the end of the secured data
Integrity Value	RC/CC/DS	Variable	The value of the RC, CS, or DS associated with the secured data

Security Parameter Indicator (SPI)

The SPI says what cryptographic operations were performed on the secured data before it was sent and thus is a guide to how to undo or otherwise take advantage of those operations.

The first byte of the SPI breaks down as shown in Figure 9-2, and the second byte of the SPI breaks down as shown in Figure 9-3, where PoR stands for Proof of Receipt.

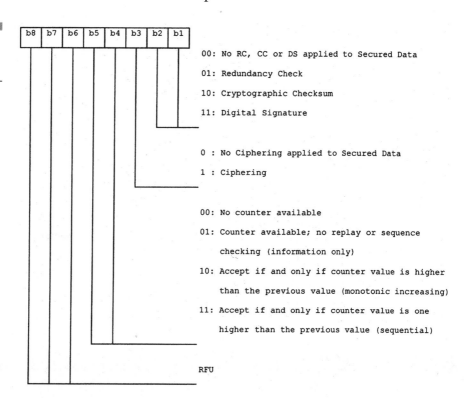

Figure 9-2
Fields of first byte of the SPI.

The encoding of the PoR is discussed in our description of the response to a ETSI TM 03.48-encapsulated SMS message.

Figure 9-3
Fields of the second byte of the SPI.

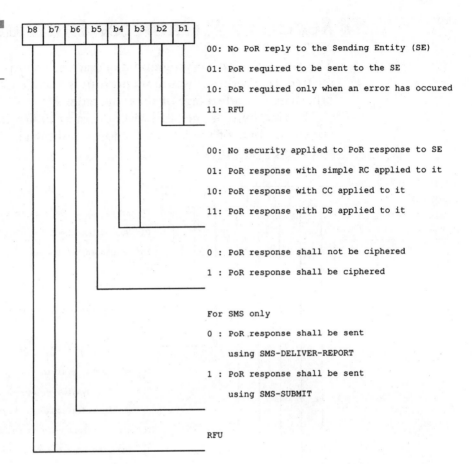

Ciphering Key Identifier (KIc) and the Key Identifier (KID)

The KIc is coded as shown in Figure 9-4 and the KID is coded as shown in Figure 9-5.

Notice that in both cases, even if the algorithm is explicitly called out, the key reference is a piece of information known only to the two ends of the communication. "Use Key #3. Wink, wink, nudge, nudge, you know what I mean."

End-to-End Security for SMS Messages

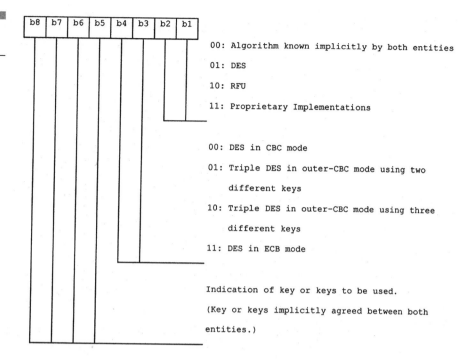

Figure 9-4
Fields of the KIc.

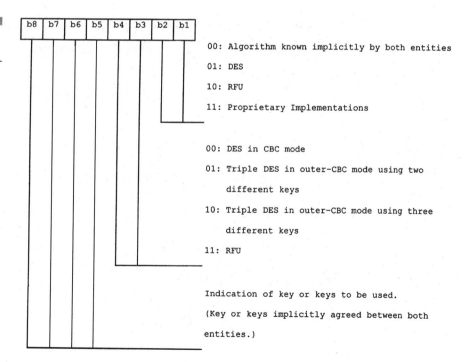

Figure 9-5
Fields of the KID.

Toolkit Application Reference (TAR)

A TAR number is a 3-byte value that has a very fuzzy definition. The official description of the TAR is "coding is application dependent." This surely is a strange definition for a field with the name "Toolkit Application Reference." How can it be application dependent if we have to decode it to figure out which application it is referencing? The result of this logical Mobius strip is that the TAR field is (ab)used by many different people for many different reasons.

The ETSI TS 101.220 standard produced by the ETSI SCP project is trying to untie this knot and has set forth the general categorization of TAR values as shown in Table 9-1.

TABLE 9-1 General Categorization of TAR Values

TAR	Application Category
'00 00 00'	Card Manager
'00 00 01' to 'AF FF FF'	Allocated by card issuer
'B0 00 00'	UICC shared file system management
'B0 00 01'	USIM file system management
'B0 00 03' to 'B0 FF FF'	Other remote file systems
'B1 00 00' to 'B1 FF FF'	Payment application
'B2 00 00' to 'BF FE FF'	RFU
'BF FF 00' to 'BF FF FF'	Proprietary toolkit application
'C0 00 00' to 'FF FF FF'	Allocated by card issuer

The TAR value was intended to serve the same function as the port field of the Internet's Transmission Control Protocol (TCP). Once the message arrives at its physical destination, the TAR or port field tells the operating system on the arrival system which application on that system will handle the message.

Because it went through a fuzzy definition phase, the TAR has been commandeered for purposes that are well outside its intended use. The USAT Interpreter standards, for example, try to use it for routing on the network. How this meshes with values in Table 9-1 is anybody's guess.

Counter (CNTR)

The counter field indexes all the messages between the application server and the SIM. Its intended purpose is to counter (no pun intended) a replay attack when somebody snags one of the SMSs and sends it again to its original destination at a later point in time. If the recipient had no way of telling that it had seen this message before, it would process it again and this could damage the overall application.

If counter management is turned on, then there are weak and strong forms of counter control. In the weak form, each received message must have a counter value greater than the immediately preceding message. In the strong form, each received message must have a counter value exactly one greater than the immediately preceding message.

You also have to keep in mind that a counter is a tiny piece of shared state with another entity, namely the entity with which you are communicating. If you are communicating with more than one entity, then you will have to maintain a separate replay counter for the other entities.

A weaker but easier way to manage the Counter field is to turn off counter management and just put a time stamp of some sort into the Counter field.

Padding Counter (PCNTR)

Many cryptographic algorithms work with blocks of bytes of a specific size. Data Encryption Standard (DES) for example, works with 8-byte blocks. If the data to be secured isn't a multiple of 8 bytes, then padding bytes, usually zeros, have to be added to the end to make the data come out to an even number of 8-byte blocks. The PCNTR says how many bytes must be added to the end of the data to make it simpatico with the algorithm(s) used to process it. Without PCNTR, we wouldn't know where the real data stopped and the padding began. Yes, that's right, the real message could end in a bunch of zeros.

Redundancy Check (RC), Cryptographic Checksum (CC), or Digital Signature (DS)

This field is used to ensure that the SMS wasn't changed along the way from the application server to the SIM and, in the case of DS, to authenticate its source. The length of the field depends on the particular cryptographic algorithm used to compute it and thus is determined by the coding of KID. As you can see, there aren't a whole lot of different algorithms called out by KID, so the usual case is "algorithm known implicitly by both entities."

Secured SMS Message Example

Let's take an up-close-and-personal look at an ETSI TS 03.48-encoded SMS message. The mandatory header is just as we described it in Chapter 3. The User Data Header Indicator bit in the first byte of the mandatory header is set to 1 because there is a User Data Header containing optional feature descriptions right after the mandatory header. The User Data Header contains at least the optional feature descriptor for SIM Toolkit Security, namely 0x70 0x00. This optional feature descriptor says that the first part of the payload is a Command Header. Of course, there could be other optional feature descriptors in the User Data Header, but for the purpose of our example let's assume that this is the only one and that all the user data—SIM Toolkit Security optional feature descriptor, Command Header, and secured data—fits in that SMS message.

Table 9-2 shows the entire User Data portion of our SMS message as sent (SMS_SUBMIT) or received (SMS_DELIVER).

End-to-End Security for SMS Messages

TABLE 9-2

User Data Portion of SMS Messages Sent and Received

Byte Number	Byte Value	Comments
1	0x02	Length of the value immediately following the User Data Header
2	0x70	Tag—SIM Toolkit Security optional feature descriptor namely user data consisting of a GSM 03.48 command packet
3	0x00	Length—the value following the SIM Toolkit Security TLV is 0 bytes long
4—5	0x001E	The entire GSM 03.48 command packet is 30 bytes long
6	0x12	The Command Header of the command packet is 18 bytes long
7—8	0x0D 0x35	SPI—redundancy check, ciphering, and counter available for information only (i.e., time stamp) on an incoming message; PoR is required with RC and ciphering sent by SMS_SUBMIT
9	0xD0	KIc—algorithm known implicitly by both entities using key 13
10	0x0x00	KID—algorithm known implicitly by both entities and uses key 0
11—13	0x00 0x00 0x05	TAR—gives the SMS to application 5
14—18	0x01 0x33 0x0A 0x0F 0x00	CNTR—Message is time stamped at 10:15:00 hours on March 3, 2001
19	0x00	PCNTR—there are 0 bytes of padding on the message
20—23	0x43d44275	RC—the value of the RC on the unenciphered data
24—35	0x55 0x72 0x79 0x79 0x62 0x2C 0x20 0x6A 0x62 0x65 0x79 0x71	"Uryyb, jbeyq"—the ROT13 enciphering of "Hello, world"

The RC was computed as follows:

```
unsigned long crc       = 0xFFFFFFFF;
unsigned char *message = "Hello, world";

main()
{
  unsigned char i, *p = message;

  do {
    crc ^= *p << 24;
    for(i = 0; i <=7; i++) {
      crc = crc & 0x80000000 ? (crc << 1) ^ 0x04C11DB7
                             : crc << 1;
    }
  } while (*++p);
}
```

This is the ITU-TSS 32-bit cyclic redundancy check (CRC).

The recipient of this SMS would look first at the SPI and conclude that the message is enciphered and comes with an RC. This means that the message has to be deciphered and then checked for redundancy.

The recipient would then look at the KIc and apply ROT13 to the secured data to reveal the plaintext of the message. The ROT is the "algorithm known implicitly by both entities" and 13 is the key (amount of rotation) used in the ROT algorithm, i.e. ROT13.

The recipient would then apply the RC "algorithm known implicitly by both entities," namely ITU-TSS 32-bit CRC, to the plaintext revealed above and get 0x43d44275. They would then compare those 4 bytes with the four RC bytes (20—23) in the Command Header, and because they are identical, they would conclude that the message hadn't been tampered with on its journey from the sender.

Now, although ITU-TSS CRC-32 is a reasonable RC algorithm, ROT13 is not a reasonable encryption algorithm, but you get the idea.

Proof of Receipt

The Command Header in our example requested that a PoR be sent back to the sender of the SMS and that this PoR should be ciphered, redundancy checked, and sent via another SMS message, i.e., an SMS_SUBMIT.

End-to-End Security for SMS Messages

The Response Packet structure is simpler than the Command Packet structure because it uses exactly the same key references as the Command Packet. Like the Command Packet, however, it includes an optional feature description in the User Data Header that flags the fact that the user data is in fact a Response Packet:

Byte Number	Byte Value	Comments
1	0x71	Tag—response packet TLV
2	0x00	Length—no Value field associated with this TLV

The fields in the Response Header are the last four fields of the Command Header plus a status code field just before the variable RC/CC/DS field that says what happened when the incoming message was processed. Table 9-3 shows the Response Header, and the possible Response Status Codes are shown in Table 9-4.

TABLE 9-3 Response Header

Field Name	Field Symbol	Length in Bytes	Description
Toolkit Application Reference	TAR	3	Copy of the TAR value in the Command Packet to which this is the response
Counter	CNTR	5	Copy of the contents of the CNTR field in the Command Packet to which this is the response
Padding Counter	PCNTR	1	The number of padding bytes at the end of the additional response data (if any)
Response Status Code	RSC	1	Disposition of message to which this is the response
Integrity Value	RC/CC/DS	Variable	The value of the RC, CS, or DS associated with the following additional response data (if any)

TABLE 9-4

Response Status Codes

Response Status Code	Meaning
0x00	Everything went swimmingly
0x01	RC/CC/DS check failed
0x02	Counter field was too low
0x03	Counter field was too high
0x04	Counter is blocked
0x05	Ciphering error
0x06	Unidentified security error
0x07	Insufficient memory to process message
0x08	Receiving application is still working on the message
0x09	TAR unknown

Let's compose an example response to the "Hello, world" message. Because the incoming message asked for an SMS_SUBMIT response, we would follow the instructions in Chapter 3 to create an SMS_SUBMIT mandatory header in which we would set the UHDI flag to 1 to indicate the presence of a User Data Header. We'd pick up the phone number of the entity that sent us the message from the SMS_DELIVER and use that as the destination number of our SMS_SUBMIT. The bytes following this mandatory header would then look like the information presented in Table 9-5.

Pairing a Sent Message with its Response

A problem that comes up again and again in building SMS systems is associating response messages with the messages they are responding to. If you have only one message outstanding to any particular mobile at a time, then this is no problem because you can use the mobile number to associate the response to the message sent. If there are multiple applications on the SIM and there are several outstanding messages to more than one application at a time, then it becomes problematic to know which outstanding message goes with a particular response.

End-to-End Security for SMS Messages

TABLE 9-5

Bytes Following the SMS_SUBMIT Mandatory Header

Byte Number	Byte Value	Comments
1	0x02	Length of the immediately following User Data Header
2	0x71	Tag—SIM Toolkit Security optional feature descriptor, namely user data consisting of a GSM 03.48 response packet
3	0x00	Length—the value following the SIM Toolkit Security TLV is 0 bytes long
4—5	0x0014	The entire GSM 03.48 response packet is 20 bytes long
6	0x0E	The Response header of the response packet is 14 bytes long
7—9	0x00 0x00 0x05	TAR—response is from application 5
10—14	0x01 0x03 0x0A 0x0F 0x00	CNTR—exactly the same Counter field in the received SMS
15	0x00	PCNTR—there are 0 bytes of padding on the message
16	0x00	Response Status Code (RSC) — everything went fine
17—20	0xe5ab9b6d	RC—the value of the redundancy check on the unenciphered additional response data
21—26	0x55 0x72 0x79 0x79 0x62 0x2C	"Uryyb"—the ROT13 enciphering of "Hello"

At first, one might think of using the Message Reference (TP-MR) field in the SMS-SUBMIT header. Unfortunately, this is not end to end, i.e., it is not carried through to the SMS-DELIVER. Message Reference is used to refer to the message as it is being held by the SMSC. For example, if you wanted to delete a message on the SMSC before it got delivered to the mobile, you would use the Message Reference number in an SMS-COMMAND message to the SMSC.

One solution to this problem is to use a ETSI TS 03.48 envelope and harness the Counter field in "information only" mode to pair messages with their responses. You'll notice that the Counter field in the response packet is mandated to be a copy of the Counter field in the command packet to which the response packet is a response.

This solution has a cost. It adds 19 bytes to the outgoing message, only 4 are used for the purpose at hand, and it adds 16 bytes to the response, 2 of which are for the status code and 4 are for the purpose at hand.

Summary

In this chapter we showed you how to build a two-way, secure, end-to-end connection between the application server side of an application and the mobile side of that application with plain old SMS messages. You still have all the benefits of the air link security provided by the underlying GSM or 3G system, including continuous challenge/response—based authentication and session key—based encryption. But air link security protects only your message, as its name indicates, when it is in the air. The GSM and 3G systems provide no protection for your message when it is zipping around the core network or waiting in a queue at one or another SMSC, or when it is sitting in the memory of the handset.

ETSI TS 03.48 lets you build your own security using your own keys and your own cryptographic algorithms so that nobody including the network operator can read or tamper with your messages.

So we have a secure link to the mobile and we have the SAT to interact with the user at the far end. What's missing is a connection between the SMS message and the Toolkit. That's what the next chapter is about.

CHAPTER 10

The SmartTrust Microbrowser and the 3GPP USAT Interpreter

All that remains is to connect incoming SMS messages with the SAT. Once we do this, we have a direct pipeline from the application server to the on-the-go mobile user and vice versa.

A program running on the SIM makes this connection. The program catches the SMSs, places calls on the SAT, and returns the results. The program can be a general-purpose or a special-purpose program that is dedicated to your particular application.

Some More SIM Toolkit History

A critical element of making this connection is being able to tell the difference between an SMS message that is intended to be read by the subscriber and an SMS message that is meant to be treated as input to a SIM-based application. In the early 1990s Swisscom solved this problem with a system called NATELsicap that enabled them to update options and routing tables on subscribers' SIMs with the use of SMS messages. A Sicap message modified the SMS header with a bit that said that this message was to be given to the Swisscom application on the SIM rather than stored in the SMS messages file. As we saw in Chapter 3, the Sicap bit was subsequently folded into the standard Data Coding Scheme (DCS) field.

Initially, "sicap" stood for "SIm Card Application Platform," but as the general utility of this technique became recognized and the number of services supported by this platform grew, particularly in the area of prepaid services, the definition of the acronym changed to "Solutions for Innovative Communications Applications." As of 2001, Sicap is a freestanding commercial enterprise and even has own Web site www.sicap.com.

At about the same time, the ETSI SMG9 standards committee was extracting itself from a run-in with the European Commission. SMG9 had been trying to define a standard way of undoing the locking of a SIM to a particular handset. SIM locking is done so that a handset whose purchase price is subsidized by a network operator can't be used with another operator's SIM. The process of unlocking a SIM using OTA commands is called "unlatching." When the European Commission got wind of SMG9's effort to standardize what appeared to them to be uncompetitive behavior, they were, as we say, not pleased. In a very deft piece of downfield running, SMG9 morphed unlatching into a general-purpose programming interface on the SIM that could be used for unlatching but certainly was intended to be used for much more user-friendly applications.

The initial enthusiasm for downloading applications to the SIM, although driven by some very elegant technical innovations such as the SIM Toolkit API discussed in Chapter 7, was soon dampened by the daunting task of securely loading applications onto the SIM and then administering those applications. As a result, there was a bit of a retreat to the original NATELsicap model of having a fixed program on the SIM to which one sends messages that are acted on and then discarded.

In 1998 a Swedish company named Across Wireless made the leap between the Swisscom NATELsicap model of mobile applications and the way that the World Wide Web works. Rather than creating proprietary SMS messages that are sent to the SIM, why not express the messages as HTML pages and think of the NATELsicap client on the SIM as a microbrowser?

In a sense, this is carrying Kristian Woodsend's original idea of a SIM Toolkit one step further. Kristian defined a standard, fixed set of commands to be implemented on the handset so that applications could be written on the SIM that didn't require any new software to be installed in the handset. A natural next step taken by Across Wireless was to define a fixed set of commands to be implemented on the SIM so that applications could be written back on network servers that didn't require any new software to be installed in the SIM. What goes around, comes around. This is exactly the NATELsicap approach.

In big iron computing, commands that are interpreted by a fixed program are called byte codes and the fixed program that interprets these commands is called, imaginatively, a byte code interpreter or, more grandiosely, a virtual machine. The only difference between the downloaded program version of SIM Toolkit applications that ran into administrative problems and the Across Wireless microbrowser version of SIM Toolkit applications is whether or not the byte codes remain on the SIM after they have been interpreted. It is little wonder then that the Across Wireless provided an attractive alternative approach where the downloaded program approach failed. It was exactly the installing of the byte codes on the SIM that caused the network operators big time headaches and sleepless nights. The Across Wireless solution provided all the flexibility and utility of general-purpose application development without the problems of SIM installation.

A Short History of Byte Code Interpreters on Smart Cards

The first smart-card-based byte code interpreters were implemented in the early 1990s. Pierre Paradinas working with Edouard Gordons and Georges Grimonprez at Lille University in France and Jelte van der Hoek and Jurjen Bos at DigiCash in The Netherlands independently and concurrently had the idea of implementing a byte code interpreter on a smart card.

In the mid 1990s Eduard de Jong implemented a Lisp-like virtual machine, Tony Guilfoyle implemented a Basic interpreter, and Keycorp and Europay implemented Forth-like interpreters. At the same time David Everett led a team at National Westminster Bank in London that started with David Watts' Pascal byte codes and ended with the Multos virtual machine.

In the late 1990s Schlumberger created a virtual machine for a subset of Java on a smart card and Microsoft came to market with a language-independent virtual machine on Smart Card for Windows. These milestones are summarized in Table 10-1.

TABLE 10-1
History of Interpreters on Smart Cards

System	Year	Team	Organization	Language(s)
CAVIMA	1990	Gordons, Grimonprez, and Paradinas	RD2P	C-Card (C)
J-Code	1990	Van der Hoek and Bos	Digicash	
	1992	Gordons, Grimonprez, and Paradinas	Gemplus	
ICEcard	1993	Knobloch	University of Karlsruhe	Treff (Forth)
	1993	Salge	Zeitcontrol	ZC-Easy (C and Pascal)
OTA	1993	Heyns and Johannes	Europay	Forth

continued on next page

TABLE 10-1

History of Interpreters on Smart Cards (continued)

System	Year	Team	Organization	Language(s)
Multos	1994	Peacham and Simmons	National Westminster	MEL (C, Java, and Basic)
OCA	1995	De Vijt and Brouillet	Europay	
TOSCA	1995	de Jong	Integrity Arts	SCIL/Clasp
HOST	1995	Cesaire, Perrot, and Richard	Oberthur	HOST(C and Basic)
CoMbO	1995	Vandewalle, Biget, and George	Gemplus	Forth
OSSCA	1995	Vuletic	Keycorp	Forth
Java Card™	1996	Wilkinson, Guthery, and Krishna	Schlumberger	Java
BasicCard	1996	Guilfoyle	ZeitControl	Basic
Windows for Smartcards	1997	Odinak	Microsoft	Visual Basic, Java and C
Wireless Internet Browser	1998	Thorstensson, Lundh, and Sellin	AU-Systems	MML, HTML and WML

When the SAT came along, it was natural to think about a scripting language for it and in fact work on an interpreter for this language started in the autumn of 1995 in the SIM department of AU-Systems in Sweden. In early 1996, an input paper was presented to SMG9 #8 that was held in Saariselkä from March 5 to 8, 1996. The paper included a proposal for the byte codes that would be executed on the SIM and proposed that they be standardized by ETSI. Here is the note from that meeting regarding this input paper:

> [T]he concept of a macro language (doc. 036/96) was discussed. The Splinter Group considered that it is not feasible to implement an interpreter on the SIM or the ME, but a compiler in the SMS Services Center (SMS-SC) for data download to different types of SIMs may be useful.

SIM manufacturers mostly were of this opinion and because they were in fact working on SIM interpreters at the time, it isn't too hard to see what was really going on. It is also interesting to note that the SIM manufacturers weren't interested in standardizing the communication between the SMSC and the SIM but rather wanted to make this channel proprietary to each SIM vendor. This would lock the SIMs to a particular vendor's OTA server and thus lock the network operator to that SIM vendor. This strategy is still in evidence today as the SIM manufacturers propose a variety of homegrown transport protocols for moving data between the SIM and the Internet.

As an aside, the representative to SMG9 from AU-System was Jonas Branden who later joined Ericsson, and he was one of the founders of WAP. Obviously, he believed that an interpreter could run on the handset.

In late 1997, AU-Systems started a Mobile Commerce project with Ericsson. It was decided that the first technology to use was SIM Toolkit because that was the only technology that could offer the required security. AU-Systems at that time was looking for a PKI-based solution, but the chips of the day didn't have modular arithmetic coprocessors. When doing the first demonstration application, AU-Systems and Ericsson understandably wanted to avoid the SIM vendor's proprietary techniques for implementing the m-commerce applications.

The application development model of the era was then (and still is in some quarters) that the bank's IT staff would do the programming. SIM manufacturers tried to teach these programmers the intricacies of their particular SIMs, not wanting to make application development too easy or too portable lest it escape into the independent application development community.

Tommy Thorsstensom, Per Lundh, and Lars-Erik Sellin of AU-System Mobile took a different approach and designed a generic application on the SIM that used HTTP and HTML as the external interface.

The initial microbrowser prototype was built on the first-generation Schlumberger 8K Java Card in spring 1998. It had an HTML external interface and the first wireless Internet gateway (WIG) byte code translator, and it supported pull and push modes of operation. But the Java Card proved too slow and provided too little RAM for the microbrowser, so a second prototype was built by Daniel Ericsson of Across Wireless on a 16K Setec card. Daniel wrote the microbrowser in byte code assembly language for the Setec virtual machine. The push mode at the time was easier to use because it wasn't based on WAP.

The SmartTrust Microbrowser and the 3GPP USAT Interpreter

The prototype became a product and has enjoyed excellent success in the market place. More than 15 different SIM card technology providers have implemented it. Telnor was the first operator to use the new browser and FilmWeb was the first commercial service. As of the middle of 2001 there are more than 50 network operators using the system and more than 8 million SIMs in the field that contain the microbrowser. Since then, many new SAT commands and PKI plug-ins have been added. It has also become much more aligned with the WAP specifications.

All of this is not to say that adding code to the SIM is not a viable way to build applications. In fact, the microbrowser is an application, so adding code is clearly appropriate. If the application is used heavily or involves high security or is focused on a particular set of users, then putting application-specific code on the SIM makes perfect sense, and we will cover SIM-based application code in the following chapters. In this chapter, we concentrate on an easier approach, the SIM microbrowser or, as the standard-compliant version is coming to be called, the USAT Interpreter.

As of mid 2001, there are three SIM microbrowsers: the original one from Across Wireless (now called Sonera Smarttrust), one originally developed by Gemplus and marketed by all the SIM card manufacturers under the umbrella of the SIMalliance, and another one called the USAT Interpreter that was wending its way through the 3GPP standardization process.

The 3GPP USAT Interpreter originally was supposed to be a merge of the Across Wireless microbrowser and the SIMalliance microbrowser but it has found a voice of its own in the process. It's not too far from the truth to say that it combines some of the best features of its two parents.

The operation of the microbrowser is virtually identical to a Web browser handling Java Script on a Web page. The incoming SMS message ("page") contains a small byte-coded program that is executed by the microbrowser. When the program terminates, the message is deleted, just like a Web page. The only difference is technical. In the case of the SIM microbrowser, the pages are reduced to byte codes for efficient transport over the air interface. This compilation takes place on a server in the mobile network called a Wireless Internet Gateway (WIG).

Originally, the SIM microbrowser was built to compete with the WAP microbrowser on the handset, and the focus was on compiling pages written in a kind of pidgin version of Wireless Markup Language (WML) into the microbrowser byte codes. As it became clear that WML wasn't going to survive and the SIM processor really couldn't compete with the handset processor, a number of different approaches were tried, which included regarding the microbrowser as merely the mobile end of a remote procedure call to the SAT.

In this chapter we examine the Across Wireless microbrowser, which is called the *Wireless Internet Browser* (WIB), and conclude by discussing the USAT Interpreter.

Sonera SmartTrust WIB

The WIB communicates with two different programs on the WIG. With respect to the first program, the WIB is a client and the program on the WIG is its server. The WIB sends ordinary URLs to the WIG server, which retrieves what the URL points to from the Internet, byte codes it, and sends the result back to the WIB. With respect to the second, the WIB acts like a server. The WIG program pushes requests to the WIB and the WIB responds.

The WIB and the WIG use the TAR value in the ETSI TS 03.48 protocol header to keep track of which WIG program is on the line. By default, if the WIB receives a message with a TAR value of 1, then it came from the WIG server (pull); if it receives a message with a TAR value of 2, then it came from the WIG client (push). It's kind of a push-me—pull-you, Dr. Jekyl and Mr. Hyde relationship (Figure 10-1).

Let's consider the situation in which the WIB is the client and the WIG is the server. Imagine for a moment that there is a menu on the screen and one of the entries looks something like this:

```
a)...
b)...
c)Hello, world
d)...
```

Selecting this menu entry causes the WIB to send the following URL to the server on the WIG:

```
http://www.acme.com/mobileapps/hello.wml
```

The SmartTrust Microbrowser and the 3GPP USAT Interpreter

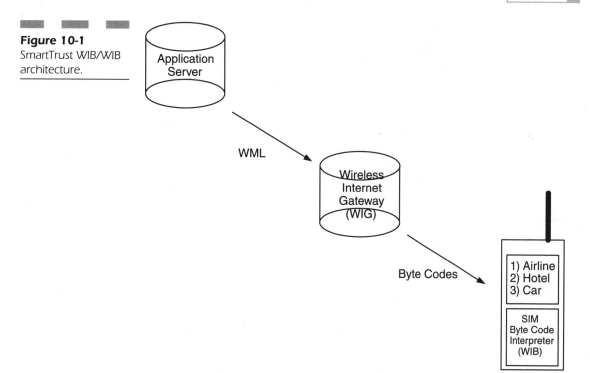

Figure 10-1
SmartTrust WIB/WIB architecture.

The WIB constructs the 03.48 envelope with the server TAR, sets the payload to the URL, and sends the whole mess off to the WIG using the SEND SMS proactive command.

The WIG unwraps the SMS and examines the TAR value. The value matches the WIG's server personality and, using HTTP, reaches out to www.acme.com and pulls back the file hello.wml from the mobileapps directory.

Let's suppose this file is simply:

```
<wml>
  <card>
    Hello, world
  </card>
</wml>
```

The WIG translates this into WIB byte codes

0012020C48656C6C6F2C20776F726C6406000600

plops them in a second 03.48 envelope with the server TAR, and sends the thusly encoded file back to the SIM in an SMS. The SIM passes the byte codes onto the WIB and the WIB follows instructions, namely displays "Hello, world" on the display of the handset using the DISPLAY TEXT proactive command.

Before taking a closer look at the SMS traffic, let's consider the byte codes. This is the one element of SMS traffic that we haven't seen before.

The first two bytes comprise the total length of the byte code encoding what follows. The byte coding itself is a sequence of TLVs with the tags shown in Table 10-2.

TABLE 10-2

SmartTrust Microbrowser Byte Codes

Byte Code Name	Byte Code Tag	Description
SUBMIT	0x01	Submit a server-bound short message
DISPLAY TEXT	0x02	Execute the DISPLAY TEXT proactive command
GET INPUT	0x03	Execute the GET INPUT proactive command
SELECT ITEM	0x04	Execute the SELECT ITEM proactive command
SKIP/GOTO	0x05	Branch to a point in the byte code stream
EXIT	0x06	Terminate interpretation
PLUG-IN	0x07	Execute a microbrowser plug-in
SET	0x08	Set a variable to a value
PROVIDE LOCAL INFORMATION	0x09	Execute the PROVIDE LOCAL INFORMATION proactive command
PLAY TONE	0x0A	Execute the PLAY TONE proactive command
SET UP IDLE MODE TEXT	0x0B	Execute the SET UP IDLE MODE TEXT proactive command
REFRESH	0x0C	Execute the REFRESH proactive command
SET UP CALL	0x0D	Execute the SET UP CALL proactive command

continued on next page

TABLE 10-2

SmartTrust Microbrowser Byte Codes (continued)

Byte Code Name	Byte Code Tag	Description
ASSIGN VERSION INFORMATION TO VARIABLE	0x0E	Read the version information EF and assign the contents to the specified variable
ASSIGN SCRIPT BUFFER SIZE TO VARIABLE	0x0F	Assign the current script buffer size of the browser to the specified variable
NEW CONTEXT	0x10	Clear the variable value table
SET RETURN TAR VALUE	0x11	Set the TAR value that the next SUBMIT command will use
SEND USSD	0x12	Execute the SEND USSD proactive command
SEND TEXT SM	0x13	Execute the SEND SHORT MESSAGE proactive command

Referring back to the encoding of the hello.wml file, you can see that the encoding is nothing more than a DISPLAY TEXT TLV followed by two EXIT TLVs. I guess the WIG wanted to make sure the WIB stopped on a dime. (Actually, one EXIT is due to the </card> and the other to the </wml>.) The DISPLAY TEXT TLV simply has the data field of "Hello, world."

Now just to get it all together, let's take a look at what comes over the SIM's transom. The WIG sends a 03.48 encapsulated SMS to the mobile containing those byte codes with the instruction to download it to the SIM. Here's what the handset sends the SIM:

```
80 C2 00 00 47 D1 45 82 02 83 81 86 07 91 91 71 09 57 22
F0 8B 36 44 03 81 11 F2 7F F6 00 00 00 00 00 00 00 27 02
70 00 00 22 0D 00 04 00 00 00 00 01 00 00 00 00 00 00 00
12 02 0C 48 65 6C 6C 6F 2C 20 77 6F 72 6C 64 06 00 06 00
```

Woof! I'm glad the SIM didn't ask directions to the train station. This has to be one of the most complicated ways to say "Hello, world" that you've ever run into.

It will be educational to deconstruct this message to the SIM. First, let's divide it into its principal parts:

```
ENVELOPE APDU
80 C2 00 00 47

SMS-PP Data Download
D1 45 82 02 83 81 86 07 91 91 71 09 57 22 F0

SMS-DELIVER TPDU
8B 36 44 03 81 11 F2 7F F6 00 00 00 00 00 00 00 27 02

User Data
70 00

03.48 Header
00 22 0D 00 04 00 00 00 00 01 00 00 00 00 00 00

WIB Byte Codes
00 12 02 0C 48 65 6C 6C 6F 2C 20 77 6F 72 6C 64 06 00 06
00
```

Now lets crack apart each principal part. The ENVELOPE APDU inserts the SMS message into the SIM:

```
80 C2 00 00 47 (ENVELOPE with 0x47 bytes of data)
```

The start of the 71 bytes is the SMS-PP data download TLV:

```
Data Download TLV: D1 (SMS-PP Data Download tag)
Total Length: 45
Device Identity TLV: 82 02 83 81 (Network to UICC)
Address TLV: 86 07 91 91 71 09 57 22 F0 (1 917 907 5220)
```

Next, the SMS Data Download TLV tells the SIM that this is an SMS-PP data download:

```
SMS Download Tag and Length: 8B 36
```

The SmartTrust Microbrowser and the 3GPP USAT Interpreter

The SMS-DELIVER header comes next. Examination of this by the handset is what led it to send the entire message along untouched by the handset:

```
Message Type Indicator: 44 (0100 0100 = SMS-DELIVER/No
                                        More Messages/
                                        User Data

Originating Address: 03 81 11 F2 (81 = 1000 0001 =
                                       Unknown/ISDN)
```

This telephone number is a little strange. You'd normally expect the telephone number of the sender here, but the sender of this SMS message is the WIG and the WIG is a server inside the VoiceStream network. Within this network, the sender is known as 112:

```
Protocol Identifier: 7F (01 111111 = SIM Download)

Data Coding Scheme:  F6 (1111 0110 = 8-bit data/SIM-
                                     specific message)

SMSC Time Stamp: 00 00 00 00 00 00 00

Total Length of User Data:  0x27 (39)
```

Next comes the User Data Header that was signaled by the UHDI bit being set in the Message Type Indicator:

```
Total Length of User Data Headers: 02
User Data: 70 00 (03.48 Command Packet)
```

Now we're finally into the payload, at least from the point of view of the SMS header. But the one optional feature description in the User Data header gives us a heads-up that the payload started with a 03.48 Command Header:

```
Length of Command Packet: 00 22
Length of Command Header: 0D
SPI:                      00 04 (Ciphering)
KIc:                      00
KID:                      00
```

```
TAR:                    00 00 01
Counter:                00 00 00 00 00
Padding:                00
```

The payload inside the last envelope contains the WIB byte codes:

```
00 12  (18 bytes follow)
02 0C 48 65 6C 6C 6F 2C 20 77 6F 72 6C 64
                        (DISPLAY TEXT "Hello, world")
06 00 EXIT
06 00 EXIT
```

You just keep reading the instructions on the outside of each envelope, opening the envelope, and doing what the instructions tell you to.

Obviously, printing "Hello, world" on the screen is no more of a killer mobile app than free calls to Romania on Tuesday. What if the menu item read:

```
a) ...
b) ...
c) Real-Time Ticket
d) ...
```

You are sitting at Gate 23 waiting for the flight to Chicago O'Hare and they've just hung out the dreaded "Mechanical Delay" sign. You take your trusty GSM phone out of your pocket, hit the microbrowser menu button, scroll down to the third entry, and hit send. This sends off the following URL:

http://www.rtt.com/rtt.wml

which retrieves this file:

```
<wml>
<card>
Enter FROM code:
<input type="text" name=FROM maxlength="3" />
Enter TO code:
<input type="text" name=TO maxlength="3" />
<go
```

```
href="http://www.rtt.com/book.asp/f=$(FROM)&amp;t=$(TO)
/>
</card>
</wml>
```

On its way through the WIG, it gets translated to the following byte codes:

```
005903120110456E7465722046524F4D20636F64653A0310020E456E
    74657220544F20636F64653A0600012D687474703A2F2F7777772E
    7274742E636F6D2F626F6F6B2E6173702F663D800126616D703B61
    6D703B743D80020600
```

These codes are passed to the SMSC in an SMS_SUBMIT envelope addressed to your phone. The SMSC passes them onto your handset, which passes them to the SIM. The SIM operating system looks at the TAR and finds that the bits are destined for the microbrowser application. It fires up the microbrowser and passes it the bits.

The microbrowser slavishly follows the byte codes' instructions and the menu pops up on your screen. You type in BOS and ORD and hit submit. Moments later, the following message pops up on your screen:

```
Go to Gate 29
```

Real-Time Tickets has booked you on another flight to O'Hare that really is going to leave in 15 minutes and saved you $550 in the process because the seat you will be sitting in was about to go to Chicago empty and the airline would rather have you sitting in it at almost any price at that moment. All this with only four SMS messages.

Actually, we could improve on this figure by storing the little FROM/TO page on the SIM so that when you picked c) "Real-Time Ticket" rather than sending off an SMS message to get the FROM/TO screen, the microbrowser just retrieved it from a file on the SIM. Not only would this be much faster but then we would have only two SMS messages—one sending your ticket request to Real-Time Travel and their response telling you what gate to go to.

But we can do even better! Imagine if the airline with the plane sitting at Gate 29 had not one empty seat but seventeen and imagine further than it could push an instantly redeemable ticket to each of the passengers biting their fingernails at Gate 23. Do you suppose they'd fill those seventeen seats?

All of the microbrowsers including the SmartTrust WIB support a push mode and a pull mode. Rather than waiting for a request for a WML page from a phone, a Web server can just send a bunch of byte codes to the WIB and the WIB will process them just as if it had requested them. There is a little state maintenance to worry about here because the microbrowser might have an outstanding request at the same time it gets the unsolicited push so it has to be able to tell that the push isn't the response to its pull. This is accomplished by assigning a different TAR value to the push. The message still goes to the microbrowser but it uses the TAR to differentiate between push and pull messages. We can name that tune with only one SMS message:

```
Tkt to ORD Rsvd
AA Flight 456
Go to Gate 29
Immediately
```

Wouldn't this be a terrific service for a travel agency to offer? After all, they have your itinerary, they know the flight is delayed, and they know your preferences in airlines. They could generate the real-time change in your schedule automatically.

The 3GPP USAT Interpreter

If you're building a mobile service based on using SIM microbrowser capabilities, you obviously want it to work on all phones. You don't want to care about which microbrowser is on the SIM any more than you care about the brand of the handset your customer is using or, for that matter, the network operator that has your account. You just want to send off a byte-coded page in an SMS message and get a reply.

Actually this is what the network operators want, which is the reason a microbrowser standardization effort in 3GPP was launched in early 2000. The result is a set of standards defining the 3GPP USAT (USIM Application Toolkit) Interpreter (Figure 10-2).

- 3GPP TS 22.112 USAT Interpreter Requirements (stage 1)
- 3GPP TS 31.112 USAT Interpreter Architecture Description (stage 2)

The SmartTrust Microbrowser and the 3GPP USAT Interpreter

Figure 10-2
SIM architecture with USAT Interpreter.

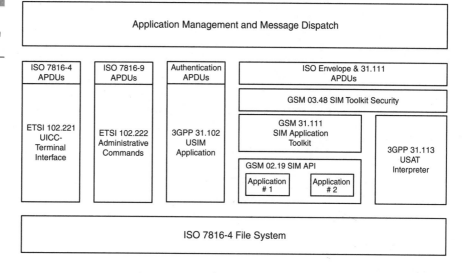

- 3GPP TS 31.113 USAT Interpreter Byte Codes (stage 3)
- 3GPP TS 31.114 USAT Interpreter Protocols and Administration (stage 3)

The SmartTrust and SIMAlliance microbrowsers were taken as the starting point for this effort because, aside from some syntax differences in the individual byte codes, they were virtually the same. The name of the SIM-resident program was changed from microbrowser to avoid confusion with the microbrowser on WAP handsets and to acknowledge that this SIM utility could be used for more than just rendering mark-up language pages.

Besides standardizing the individual byte codes, 3GPP 31.113 standardizes how the byte codes are organized in the message sent to the SIM. A diagram of this organization is shown in Figure 10-3.

The USAT Interpreter page starts with some general page parameters and then consists of a series of navigation units. Each navigation unit can be thought of as a screen of information and interaction. They are called navigation units because you move from one to another as you make choices. A navigation unit in turn starts with a name so that you can jump to it and then contains a series of byte codes, which is where the action is. The whole mess is, of course, encoded as a big Russian-doll compound TLV.

A very useful innovation that was added to the USAT Interpreter was local name spaces. The byte codes can contain constants such as

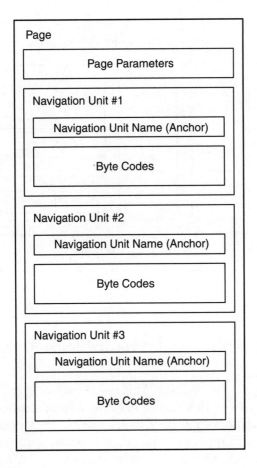

Figure 10-3
USAT Interpreter page of byte codes.

the phone number to which to reply or the size of the text that is to be displayed, but they also can refer to variables that can hold such information. Not only does this make it easy to write general-purpose USAT Interpreter pages but it also makes the SMS messages shorter, thus saving on bandwidth and transmission costs.

There are four name spaces that can hold USAT Interpreter variable values:

- **Environment Area**—System information such as Interpreter version
- **Permanent Area**—User information such as name and e-mail address
- **Temporary Area**—Scratch space for current page
- **Page String Area**—String pool for current page

The SmartTrust Microbrowser and the 3GPP USAT Interpreter

At the end of the day, there are only a dozen byte codes and two of those are optimizations (Table 10-3). Some of the commands have two tag values because the high bit is used to indicate a command modifier byte as the first byte in the value field. The second and subsequent bytes are the real value field. The length byte counts all the bytes, command modifier bytes, and value bytes.

TABLE 10-3

USAT Interpreter Byte Codes

	Tag	Description
Navigation Byte Codes		
Set Variable	0x14	Set a variable in a namespace to a value
Assign and Branch	0x15	Display a menu and branch to choose from
Go Back	0x19/0x99	Jump to the navigation unit on the top of the history list
Branch on Variable Value	0x1A	Switch statement; test variable value and branch to associated element of a list
Exit	0x1B/0x9B	Terminate interpretation of page
Bug Out Byte Codes		
Execute SAT Command	0x1C/0x9C	Place a call on the SIM Toolkit API
Execute Native Command	0x1D/0x9D	Place a call on a SIM plug-in
Data Handling Byte Codes		
Extract	0x16	Assign a span of bytes in an array to a variable
Get Length	0x1E	Assign the total length of the values of a list by matching one variable to another variable
Get TLV Value	0x1F	Set a variable equal to the value field of a TLV
Optimizations		
DISPLAY TEXT	0x20	Issue a DISPLAY TEXT SAT Command to the handset
GET INPUT	0x21	Issue a GET INPUT SAT Command to the handset

Let's take a quick up-close-and-personal look at a USAT page (Table 10-4).

TABLE 10-4

Example USAT Interpreter Page

Byte Number	Byte Value	Description
1	0x01	Tag—Page
2	0x0D	Length—length of the page
3	0x02	Tag—page identification
4	0x01	Length—length of the page identification
5	0x17	Value—this is page 23
6	0x09	Tag—navigation unit
7	0x09	Length—the length of the first (and only) navigation unit
8	0x20	Tag—DISPLAY TEXT byte code
9	0x04	Length—Length of the value field of the byte code
10	0x0D	Tag—inline value
11	0x01	Length—length of inline value
12	0x29	Variable reference—user name
13	0x9B	Tag—exit byte code—attribute byte present
14	0x01	Length
15	0x01	Attribute byte—terminate session

When this page runs through the USAT byte code interpreter, it shows the user's name on the screen and then quits. It's kind of a "Hello, myself" routine. This is not terribly useful unless you're given to forgetting who you are, but it does contain all of the elements of a USAT Interpreter page.

The good news is that a SIM-based byte code interpreter is being standardized. The bad news is that there is some noticeable lag in the process and it will probably be at least a year before any SIMs containing the USAT Interpreter start showing up in the street.

Remote Procedure Call Using the USAT Interpreter

SIM microbrowsers and the USAT Interpreter were originally thought of as being driven by mark-up languages such as WML, cHTML, and Basic XHTML. The idea, hotly denied by the SIM manufacturers, was to compete with WAP browsers on the handset. You can sense this design intent in the structure of the page and even the use of the term "page" to describe the interpretation context.

The Perform SAT Command byte code does provide general-purpose remote access to the full capabilities of the SAT. As a result, in addition to being an attractive encoding for mark-up languages, the USAT Interpreter byte codes are an attractive encoding for remote query languages such as the familiar HTML URL and even for remote procedure calls (RPCs) from standard functional languages such as C, Visual Basic, and Java.

The basic idea is to translate the URL or RPC into the byte code that places the call locally on the SIM and then to surround that byte code with byte code instructions that ensure that the call is placed efficiently and that the results of the call are returned to the program issuing the URL or RPC in a comprehensible fashion.

The syntax of the RPC on the caller's side depends on the language being used to build the program that makes the call. In functional languages, it looks like a regular function call with the proviso that you add the telephone number of the SIM on which you want the call to be executed. The traffic from a C program and RPC to the USAT Interpreter on a SIM might look like this:

```
pick = GetInput("16172301346", "To?");
```

From a Web page the same RPC in URL format might look like this:

```
http://www.telecom.com/+16172301346/getinput?text="To?"
```

where www.telecom.com is a Web server at the SMSC of the network operator or an independent mobile service provider.

In either case, the RPC is translated into two byte codes: GET INPUT, which displays the text and acquires the input from the user, and ASSIGN AND BRANCH, which returns the user input to the caller. A third byte code, GET LENGTH, is used to compute the length of the string entered by the user.

The GET INPUT byte code TLV to capture the user's key hits is shown in Table 10-5. The Branch On Variable Value byte code to return the input to the remote calling program is shown in Table 10-6. The Get Length byte code that set Variable 2 to the length of the response from the user is shown in Table 10-7.

TABLE 10-5

GET INPUT Byte Code TLV

Bytes	Value	Description
1	0x21	Tag—GET INPUT
2	0x06	Length of GET INPUT
3	0x81	Store input in temporary variable 1
4	0x0D	Tag—inline value
5	0x03	Length of inline value
6—8	0x54 0x6F 0x3F	"To?"

TABLE 10-6

Branch On Variable Value Byte Code

Bytes	Value	Description
1	0x15	Tag—Assign and Branch
2	0x28	Length of Assign and Branch
3	0x00	Destination variable (not used)
4	0x10	Tag—ordered list
5	0x26	Length of ordered list
6	0x11	Tag—page reference
7	0x24	Length of page reference
8	0x12	Tag—page parameters
9	0x22	Length of page parameters
10—37	HTTP Reply	"HTTP/1.0 200\nContent-Length:"
38—42	0xC0 0x82 0x0D 0x0A 0xC0 0x81	Length of HTTP body from Temporary Variable 2, carriage return, line feed, HTTP body containing user input from Temporary Variable 1

TABLE 10-7

Get Length Byte Code.

Bytes	Value	Description
1	0x1E	Tag—get length
2	0x04	Length of get length
3	0x82	Store result in Temporary Variable 2
4	0x0C	Tag—variable list
5	0x01	Length of variable list
6	0x81	Compute length of Temporary Variable 1

Fifty-six bytes comprised the entire RPC. Even with the 03.48 envelope, these bytes can easily fit into one SMS message.

In the URL format you could send multiple commands, separated by semicolons, which would be bundled up into one RPC page:

```
http://www.telecom.com/+16172903963/
    playTone?tone=beep;selectItem?items=(CGX, MDS, ORD)
```

The response of such a compound RPC typically would be the response of the last command executed.

Besides being useful in its own right, the RPC construct is a great way to quickly develop and debug USAT Interpreter programs that eventually will be sent over the air and run on the SIM.

Summary

In this chapter we described in detail the use of SIM-based microbrowsers and the USAT Interpreter for building mobile applications. Although they were modeled on downloaded byte-coded applications, microbrowser applications have gotten a better reception in the marketplace because they can be developed more quickly and are much easier to administer.

Most SIMs in the field, particularly in the United States, don't support a microbrowser, so you may have to work directly with the network operator to make sure that the folks you want to use your application are equipped with SIMs that contain microbrowsers. This is much easier than persuading the network operator into downloading your application to their SIMs. You can start with the SmartTrust microbrowser today and stay with it as it evolves to the USAT Interpreter in the future.

CHAPTER 11

The USAT Interpreter at Work

As the second wireless carrier in Denmark's highly competitive telecommunications market, Sonofon has a mission to capture the lead by rolling out the most advanced services to its customers. In spring 2001, Sonofon had a subscriber base of more than 900,000 users—an impressive accomplishment in a country with only 6 million in total population. To stay ahead of the wave, Sonofon early on adapted wireless data channels and tools, from SMS to SAT to WAP. But after three years as a technology pioneer, Sonofon was still searching for the ideal balance of tools to provide end users and content provider partners with easy expansion and customization of services along with high performance and security. In 2001 this wireless carrier decided that implementation of a SIM-based microbrowser, more formally known as the USAT Interpreter, provided a solution that works for most Sonofon subscribers today and creates a flexible foundation for rapid expansion and upgrading of data services in the future. With the microbrowser installed on all new SIMs issued by Sonofon, the carrier is now in a position to enhance existing services and implement the next generation of commerce.

Business Drivers

Value-added services have become the business holy grail for wireless operators around the world. The typical wireless carrier is caught in a squeeze between a decline in the price that can be charged for voice services and increasing costs for implementing next-generation network infrastructure and upgrading in-house technology. As voice services attract fewer new subscribers, the carriers look to higher-priced data services to pick up the slack and generate future profits.

To achieve this goal, however, carriers must master three difficult challenges: (1) offering services that are popular with and easily accessible to the majority of their subscribers, (2) ensuring high performance and security standards at the point of use, and (3) creating a business model and pricing level that motivates content and service partners as well as subscribers. Getting all three factors in balance has been an elusive goal for many carriers.

With an understanding of all the challenges involved, Sonofon has paid careful attention to integrating its business and technology strategies for implementing value-added services. Its management views this integration as the best path to winning new business and obtaining competitive advantage in the crowded Danish wireless market.

The USAT Interpreter at Work

Thomas Brunn Pedersen, Sonofon's SIM and Smart Card Manager, pointed out that there are multiple business drivers for value-added services in general and the microbrowser implementation in particular.

> We are aiming to achieve a number of the expected benefits that are associated with implementing value-added wireless applications. We have already seen that the most popular applications will generate increased traffic on the network, and that getting users excited about these applications will also give us an edge in attracting new customers and keeping our current customers loyal. In the longer run, high value applications will lead to increases in key measures like Average Revenue Per User [ARPU]. But we don't take these benefits for granted. We know that creating applications is just the first step—making sure that they genuinely meet the needs of the market and updating the features quickly in response to customer feedback are essential requirements for success.

The importance of keeping up with customer tastes and being able to change applications and services on a regular basis were the most important considerations in Sonofon's implementation of the microbrowser as part of its business and technology strategy. As Pedersen noted,

> The microbrowser was the most effective approach to creating a dynamic Sonofon-branded mobile portal that could meet the needs of customers today by delivering exciting, flexible services to the subscriber market right away. We wanted to keep pace with new developments and enhancements without forcing our customers to upgrade their handsets to WAP or wait for next generation networks. Implementing a SIM with a microbrowser provided us with a very easy way to implement new services and deliver them to customers in a quick way. Once the customer has a [microbrowser] enabled SIM, we can keep adding new content and services via the Sonofon Web site. Now the individual doesn't have to do a SIM card exchange or anything else that could be a barrier to using the services on a daily basis.

The ability to add new applications to SIMs that are already out in the field in customer phones provides some important advantages in working with content providers. The microbrowser is a good match for today's environment for content programming on the Web because it acts as an "interpreter" for delivering content from WML pages to the handset. Because the microbrowser supports building new customized menus on the Web and downloading them to the phone over the air, it also provides a platform for quickly implementing new services. Knowing that new features and services can reach the majority of subscribers is a real motivation for the content providers to keep expanding and upgrading their services.

Sonofon has estimated that, with some basic training in microbrowser applications, today's Web and WAP content providers will be able to provide the code themselves or at least do 75 percent of the work required to get new SIM-based applications to the subscribers' handsets. This in turn saves Sonofon time and money in application deployment and allows it to offer a much broader selection of services than if it had to develop the majority of value-added applications in-house.

The microbrowser applications can be accessed by a much larger number of subscribers today than comparable WAP-based services. Pedersen notes that Denmark has better than a 66 percent mobile penetration rate for its entire population. Of the phones that are already in subscriber hands, he estimates that almost 75 percent already have SIM Toolkit-enabled phones. That means that the microbrowser solution is readily available to a very significant percentage of current mobile users compared with the much smaller number of subscribers actually using WAP phones and services. As noted it the Technology Overview, Sonofon doesn't highlight the microbrowser technology per se in delivering services to its subscribers. Instead, it focuses on the value from the customer's point of view and then makes it as easy and fast as possible for customers to experience that value. So the microbrowser is one of a number of technologies that has been integrated to create the best possible experience for the greatest number of subscribers.

Technology Overview

Starting With SMS

The first step in Sonofon's drive toward value-added services was to capitalize on the popularity of SMS messages among many different segments of its subscriber base. In 1998 Sonofon created a new service called Gismo to offer a variety of entertainment services that were targeted at the youth audience along with financial and other information aimed at business users. As initially designed, Gismo was based on a combination of pushing SMS messages to the user handset with a predetermined menu of services loaded onto the SIM via the SAT. Users could select from this menu to activate particular services or to pull information from the Gismo portal through a query function.

The USAT Interpreter at Work

The original Gismo mobile portal offered about 50 content services, ranging from news and business information to weather, jokes, and entertainment along with subscriber-related services such as adding minutes, checking account status, or changing subscription plans. The Gismo developers used the SAT to preload menus for the services that were expected to be the most popular directly onto the SIM card. However, this selection was really just a guess based on estimates of subscribers' behavior and tastes. Figure 11-1 illustrates the categories of information that were available from the original Gismo implementation.

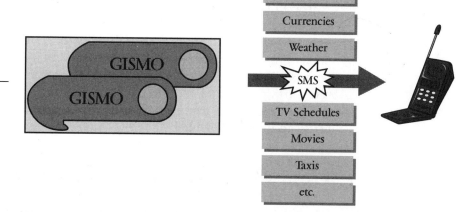

Figure 11-1
Information categories built into the early Gismo mobile services menu.

As Gismo attracted more subscribers, it very quickly sprouted a variety of new content and services based on customer feedback. After just two or three months, the portal already had more than 300 services It was becoming clear that the model of loading a static number of mobile services onto a SIM using the SAT was not the best approach for customers who wanted more dynamic, seasonal types of data services. The ideal would be to give customers a means to update services and customize the display on the handset whenever they wanted without creating extra technical or customer support overhead for Sonofon.

Pedersen noted that Gismo's success in attracting subscriber interest and the rapid growth of new types of services highlighted the limitations of the original technical design of the service.

> We started out with an older way of thinking that was based on slower growth and more stability in what we would be offering to subscribers.

This model has assumed that it would be more efficient to simply bake the most popular services directly into the SIM. The idea was that a small core list of services would serve the needs of the majority of our subscribers. We assumed that we would be able to predict pretty accurately what exact services each subscriber was most likely to want to access. But we hadn't really taken into account the different tastes of different age groups and the interest that customers would have in selecting different combinations of services that would be unique to them. So our original business model and technology was too limiting. Also, we didn't provide for the continued expansion of each service, or the proliferation of different types of programs that could be provided by business partners.

It soon became clear that keeping up with the way subscribers actually used the service and allowing them to have easy access to all the new information resources that third-party content providers were making available would require a different technology solution. Because WAP was generating lots of attention as a new standard for integrating Web and mobile content, Sonofon decided to include a WAP portal in its technology roadmap. That step, however, didn't really address the problem.

From WAP to One Integrated Portal

Like other wireless carriers, Sonofon used WAP to integrate Web content with wireless services. In 2000 the carrier created a WAP portal for Gismo and other services only to find that this didn't generate the expected boost in traffic and customers. In hindsight, Pedersen noted that one problem with Sonofon's initial WAP strategy was that it was technology driven rather than responsive to specific customer demand.

> Like most operators, we tried to get customers excited about the WAP technology itself, instead of communicating about a user-oriented service or a specific added value. Then it turned out that it was expensive to upgrade to the WAP handsets and the actual experience of using WAP to download information was pretty slow and frustrating. So it is not that surprising that the customers were not motivated to migrate to WAP. They just didn't have a clear incentive. Plus, we saw that even when customers did sign up for WAP services they weren't very likely to use them much.

Instead of waiting for future WAP improvements and gradual user acceptance, Sonofon implemented a more accessible category of wireless services that could communicate more directly with users. One step was to create a comprehensive mobile portal that integrated all the available technologies in a way that was transparent to the user. This included Web-based information, WAP, information push-and-pull services based on SMS, and the services that were being offered under the Gismo brand. The SIM-based microbrowser facilitated this integration and allowed users to select and update the level of services delivered to their phones as often as they wished. However, according to Pedersen, Sonofon didn't emphasize the integration of different technologies. Rather, they focused on getting all the different components—the Web, WAP, SMS push-and-pull services, and the SIM browser portal—to work together.

At this point, Sonofon also had to focus on the end user and make a clear value case for all the services, so they didn't try to promote the SIM microbrowser technology name or create a new brand of services. Instead, they used the Sonofon brand to facilitate the branding of the new portal. Now one Web address, www.sonofon.dk, gives access to an integrated mobile portal that is the starting point for all customers.

Figure 11-2 provides an overview of the key components of Sonofon's integrated portal strategy.

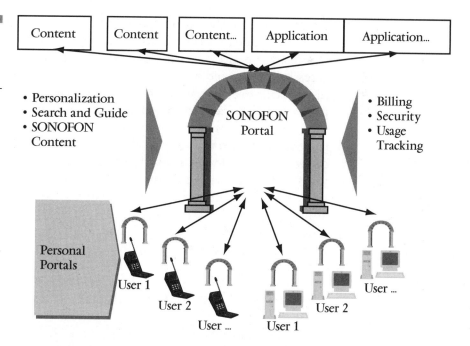

Figure 11-2
Customer orientation of the Sonofon integrated portal.

Integrating with the Microbrowser

At the same time, Gismo was still attracting lots of customers with its SMS-based information and content services, so Sonofon decided to look more closely at how to build more value and flexibility into this service. SMS was very good for messaging and push of information services in preselected categories. The SIM Toolkit worked well to load core, stable applications securely onto the phone, but it had proved to be a cumbersome approach for handling large numbers of rapidly changing, frequently updated services. The logical next step, according to Pedersen, was to add a microbrowser to the SIM. Unlike the WAP browser, the SIM-based microbrowser worked with most of the existing handsets and could handle both push and pull types of services.

> By 2001 we had learned enough from our early experience to see the limitations of relying on just one solution. We took a look at all the options and actually decided to do something wise to expand our services in an inclusive way using the microbrowser. We saw that WAP is just not mature enough to be the primary service delivery mechanism. Even when users had the right phone to handle WAP, the experience wasn't great and WAP was not generating enough traffic on our portal.

Expansion of the Gismo portal was the first step. This provided a constantly updated central location where customers could find out about all the latest available services. Because not all services were accessible from older phones and SIM cards, the Gismo portal also provided customers with information about how to get the right phone and SIM card for more advanced options.

Most customers now can add any new applications they want from the Sonofon portal over the air (OTA) onto their existing SIM. This capability encourages subscribers to sample new services and motivates content providers to keep upgrading their services. The combination of ease of use and frequently upgraded services in turn makes Sonofon services attractive to new subscribers and increases the amount of SMS messages that subscribers are likely to use, thus increasing revenues for the carrier.

Moving to Mobile Banking and M-Commerce

With the microbrowser in place and well accepted by customers, Sonofon also is well positioned to expand its m-commerce and higher end services. It is currently using the SIM browser as the basis for launching a mobile banking application. The banking application uses the microbrowser to connect Sonofon's system with Sydbank, a bank computer center for 20 banks in Denmark. The basic architecture of this mobile banking application is shown in Figure 11-3.

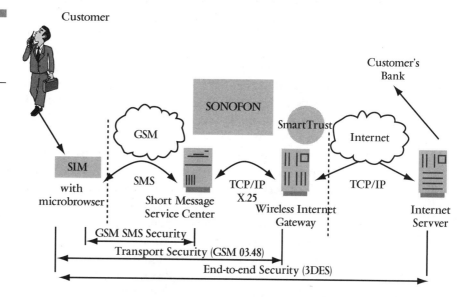

Figure 11-3
Components of the Mobile Banking Project.

One very attractive feature of using the microbrowser architecture to implement mobile banking is the fact it supports high standards for privacy and security. It provides for full, end-to-end encryption of all traffic so that Sonofon doesn't even get to see the actual transaction that takes place over its network; this is kept private between the customer and the bank The pilot of mobile banking was rolled out in spring 2001 with the leading bank within the Sydbank group. As of April 2001, the bank has signed up about 3,000 of its customers for this service. The aim is to have 140,000 subscribers for mobile banking across Denmark within three years.

From the User Point of View

Sonofon customers who want to initiate new mobile services and options on the Web or their phones can go directly to www.sonofon.dk. There a menu that guides them to specific services that they are interested in subscribing to or introduces them to some basic predetermined service packs. This interface also has some built-in intelligence. Based on subscriber information, it helps users to figure out what is available and what new services they could download directly from Sonofon. The mobile portal provides a wizard function where the user types in his or her GSM phone number and the type of phone. This information then goes back into a Sonofon database that looks up the type of SIM that this individual user currently has in his or her phone and checks with Sonofon's billing system to validate what services are available given this user's current configuration.

Once this quick backend search is complete, the mobile portal displays the current configuration of services that are available to that user. If the customer doesn't have a version of the SIM that supports access to the latest microbrowser-based content or the banking application, the portal response instantly tells the user what it will take to upgrade to obtain the desired services. If the customer has an older SIM without the microbrowser, the portal wizard will let him know what other services will become available if he is interested in upgrading. If he expresses interest, it will then provide information about how to upgrade the physical SIM. To motivate customers to upgrade, Sonofon makes this process as painless and cost effective as possible. To obtain a new SIM, the customer will pay a fee of about $12 (US) to cover the cost of the physical SIM exchange. The actual request for upgrade can be made online, and the fee is added to the phone bill. Customer service at Sonofon makes the exchange and the customer keeps the same phone number.

In keeping with their strategy of encouraging all subscribers to use value-added services, Sononfon provides some packages even for those users who don't choose to get an upgraded SIM. Even without the microbrowser in place, users can order a Mobile Menu Basic package directly from the portal.

Mobile Menu Basic offers two predesigned services packages that the user can select to download OTA and overwrite their existing services. Basically in this option, users get preconfigured packages that include different combinations of services at a set price.

These packages are downloaded via the SMS channel, and the user who switches off her phone and puts it back on has the selected service package instantly available.

A more advanced option that takes advantage of the features of the microbrowser SIM is called Mobile Menu Plus. This option tells users how to gain access to applications like Internet pages, mobile banking, travel planning, and trading services and how to customize the on-phone menus for those applications. One motivation for upgrading to the microbrowser is that the users gain access to a much broader variety of applications and they can download those applications easily in a self-service mode.

To begin subscribing to a particular service, the user can go onto the appropriate mobile portal page and drag an application onto the Web screen. It is also possible to create a customized mobile portal suited to that individual using the Web interface and then get a new menu that reflects these services downloaded OTA into the phone using the microbrowser.

Implementation Challenges and Strategies

All in all, the Sonofon portal is designed to make life as easy as possible for the subscriber who wants to add more services. Achieving this simplicity on the front end took a lot of hard work behind the scenes at Sononfon and with business partners. Sonofon engineers had to work internally to integrate the customer and billing databases that would provide dynamic information on the portal through the configuration wizard. Sonofon also had to create the initial applications that would take advantage of the microbrowser. Its long-term model, however, is to build most of its services on third-party application and content providers. This presents its own challenges.

According to Pedersen, one of the biggest requirements for Sonofon has been convincing a critical mass of content providers and key application providers to build their applications to leverage the microbrowser capabilities. Although this isn't difficult technically, it does require some education about the how the microbrowser is different from WAP and what can be reused from existing WAP applications:

> In motivating and educating the content providers and the key application providers to partner with us, we have seen a lack of understanding about the SIM browser concept. It has been a real task to convince content providers that were always thinking about WAP that the SIM toolkit is a better immediate solution that can deliver their content to the market today. We had to really demonstrate all the features that we were getting from a SIM microbrowser, and also show them that it was not a whole separate technology they had to learn from scratch.

It took Sonofon several months of educational effort before content providers were convinced to build for the microbrowser based on its superiority in handling secure transactions and the higher number of phones that could support it. Sonofon discovered that there were some key selling points that helped to convince partners to make the initial effort required to develop microbrowser applications.

> We were able to tell content providers that the SIM browser implementation would be in the same language that would be used for the future of WAP. That means that they don't have to do a completely new development for each technology. They could reuse the implementation that was being done for the microbrowser on a future WAP portal. There are some differences between doing WML pages for SIM and WAP but at least it is the same way of thinking.

Sonofon invested in this educational effort as part of its long-term strategy to sign on application developers and content providers as key business partners for its service offerings. For example, in implementing the mobile banking application, Sonofon got the application developers from the banking center involved from the very start. According to Pedersen:

> We held three workshops together with the programmers from the bank showing them a small prototype of a microbrowser application and how it handled transactions in general. That first workshop showed how the microbrowser would work and what it would look like. We spent a day using this to educate the bank's IT staff and programmers. We showed them how to do a small "hello world" application and demonstrated how to do plug-ins and use the key features of the micro-browser.

Once the bank developers had the basic tools in hand, they designed the application on their own. But Sonofon provided continued support in terms of making sure the design was optimal for the microbrowser environment. Pedersen concluded:

> After the initial session, developers then went home and did a design and wrote specifications for an application They then came back for a kind of code review where Sonofon's own designers did a critique and talked about the differences of the SMS channel from using a data channel using WAP. The developers saw that they had to think about this more carefully. Then they went home and worked with their internal application designers and got the functionality all set. At that point, they were ready to implement an application that really met their needs and was well suited to the mobile banking customers.

This approach worked well to get the mobile banking project up and running, and given the high value of the banking transactions, it is likely to pay off in a fairly short time. But Sonofon recognizes that having to do this for every application provider would end up being far too time consuming and expensive, so they are looking for better ways to educate the developer community about the easiest way to get applications up and running.

Sonofon is convinced that the microbrowser offers the best combination of fast implementation, flexibility, and ease of use for the subscriber.

> The fact that a microbrowser application is easy to change and to customize by individual users or by different service providers who may not have in-house developers gives it a big advantage over other solutions. This makes it more likely to have a number of active providers and encourages customers to try out new services.

Bottom-Line Benefits

In addition to all the long-term strategic advantages of value-added applications in terms of customer loyalty and business partnerships, Sonofon gets the immediate benefit of revenues from certain types of services. In general, there is a charge of approximately 0.05 Euros for application-related SMS messages originating from the handset. The banking application is charged at a higher rate—currently 0.1 Euro per message. This revenue model probably will expand to include some fees from the banking partners as the application is deployed more widely.

There are also a number of costs that have to be recovered for the value-added services to become profitable over the long term. Some content providers are paid commissions by Sonofon to ensure that the information in demand by subscribers is available. Other providers distribute their information for free and get back value in advertising and use the mobile service to attract customers dynamically, like a cinema that provides film information and times to advertise their current movies. As more subscribers access these services, there is more motivation for content providers to participate and add new features.

Sonofon also develops some of their own applications in-house, especially those related to subscriber information and phone-related services. To orient customers to the different services available, they use their

Web site to provide specific instructions on how to set up and use each service.

Lessons Learned

In their early drive to be a leader in value-added mobile services, Sonofon has learned a number of lessons that they are happy to share with other carriers and application developers. According to Pedersen,

> We found that the different technology we wanted to roll out was not fully mature. That meant that we ran into problems in the early development stage that we had to solve in house and at the bank. Sometimes we could not launch applications because they were residing in the bank servers. It took more time than we expected to do all the mobile portal development and integrate it with the billing system. Completing the IT background work to get the portal ready for use took about 8 months. Despite some of these early delays, we feel strongly that the SIM microbrowser is the natural choice for implementing existing and future services. It will make it easier for us to launch the next generation of mobile commerce applications. We have found that the SIM Browser is a mature platform to launch, and it is a great way to motivate people to try out a wide variety of services—this is something that is ready now, long before we can count on WAP becoming a mature platform. And that makes it a strategic choice for us.

The deployment of the microbrowser has made it easier to sign up customers for new services and for Sonofon to work with content providers to keep adding new features. At this point, the growth focus is on Mobile Menu Plus—service packages that can be downloaded directly to the phone and customized to suit individual user tastes. The Mobile Menu Basic is positioned to give existing customers a taste of applications and encourage them to upgrade to the more dynamic service options using the SIM browser.

Sonofon has accomplished its strategic goals of increasing revenue-generating SMS traffic and attracting new customers while motivating existing customers to sign on for expanded service options. It sees the microbrowser platform as an effective path to implementing more m-commerce applications such as mobile banking over the next few years.

CHAPTER 12

The USAT Virtual Machine and SIM Toolkit Programs

In Chapter 10 we built SIM applications by using a subset of a mark-up language called WML. Each WML page was sent into the network, translated into byte codes by the operator's WIG, and the byte codes were sent to the SIM using SMS messages. When the byte coded WML page got to the SIM, it was handed off to a program called the USAT Interpreter that rendered the page by executing the byte codes. During page rendering, the byte code interpreter used the SAT API to display the page on the screen of the phone and conduct various interactions with the user.

This works fine if your programming model is based on transactions, i.e., consists of a series of fire-and-forget pages with occasional user interactions as on a Web site. In fact, a surprising lot of very useful and profitable mobile applications can be built with this model, as the case study of Chapter 11 showed. There is a school of application design that says that all useful mobile applications are based on transactions. But you are probably wondering what you'd do if you wanted to write a program in a procedural language such as Visual Basic, Java, or C rather than a mark-up language such as WML, XHTML, or cHTML.

There is a technology that allows you to download full-bodied procedural language programs to the SIM. From 50,000 feet, the handling of procedural programs works a lot like the handling of mark-up language programs. Your program is converted to byte codes and sent to the SIM using SMS messaging. On the SIM, the byte codes are executed by a virtual machine. And just like mark-up programs, procedural language programs have full access to the SAT API. But as you come down to say 10,000 feet, you can start to see some differences. This chapter is about SAT programs and their handling by another byte code interpreter on the SIM, the USAT Virtual Machine.

Table 12-1 summarizes the significant similarities and differences between USAT Interpreter programs and USAT Virtual Machine programs. A difference is the byte codes that they execute. The USAT Interpreter executes a small number of byte codes that mostly concern rendering and navigating a mark-up language page. The USAT Virtual Machine has a larger number of byte codes that support evaluation of arithmetic expressions and testing and branching on logical expressions. Both support name spaces (variables with values) and a notion of subroutine calls and both provide complete access to the SAT API.

The USAT Virtual Machine and SIM Toolkit Programs

TABLE 12-1

USAT Interpreter versus USAT Virtual Machine

	USAT Interpreter	USAT Virtual Machine
Access to SAT API	Yes	Yes
Uniform human interface	Yes	No
Loading security control	Low	High
Execution security control	High	Low
Stored permanently on SIM	No	Yes
Connected to Internet	Yes	No
Access to SIM file system	No	Yes
Program execution logic	Little	Lots
Software development kits	No	Yes
Development time	Short	Long
SIM Toolkit 03.48 end-to-end security	Yes	Yes
Byte coded for efficient transmission and execution	Yes	Yes
Availability (approximate)	80 million SIMs 60 networks	10 million SIMs 5 networks

The fact that USAT Interpreter pages are transient and USAT Virtual Machine programs are permanent is more a matter of convention than technical necessity. On purely technical grounds, USAT Interpreter pages could be stored permanently on the SIM and USAT Virtual Machine programs could be executed once and then thrown away. The convention is due to the fact that USAT Virtual Machine applications tend to be larger than the collection of pages comprising a USAT Interpreter application and more caution and effort has to be exercised in loading a Virtual Machine program on a SIM as compared to a WML page (Figure 12-1).

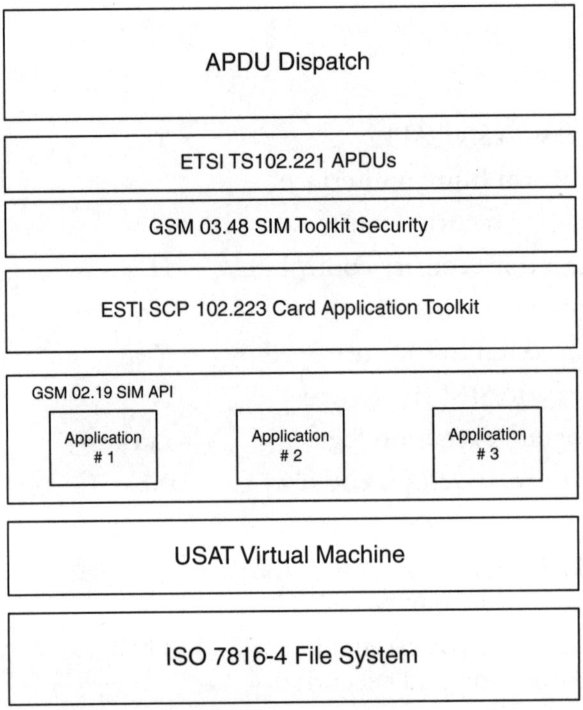

Figure 12-1
USAT Virtual Machine stack.

```
┌─────────────────────────────────────────┐
│           APDU Dispatch                 │
└─────────────────────────────────────────┘
┌─────────────────────────────────────────┐
│         ETSI TS102.221 APDUs            │
└─────────────────────────────────────────┘
┌─────────────────────────────────────────┐
│      GSM 03.48 SIM Toolkit Security     │
└─────────────────────────────────────────┘
┌─────────────────────────────────────────┐
│  ESTI SCP 102.223 Card Application Toolkit │
└─────────────────────────────────────────┘
┌─────────────────────────────────────────┐
│  GSM 02.19 SIM API                      │
│  ┌──────────┐ ┌──────────┐ ┌──────────┐ │
│  │Application│ │Application│ │Application│ │
│  │    #1    │ │    #2    │ │    #3    │ │
│  └──────────┘ └──────────┘ └──────────┘ │
└─────────────────────────────────────────┘
┌─────────────────────────────────────────┐
│         USAT Virtual Machine            │
└─────────────────────────────────────────┘
┌─────────────────────────────────────────┐
│        ISO 7816-4 File System           │
└─────────────────────────────────────────┘
```

Variants of the USAT Virtual Machine

Just like the USAT Interpreter, there are three different USAT Virtual Machines in the field. There is one based on the Java Card virtual machine that runs programs that are translated into the Java Card byte codes. These programs are typically written in the Java Card variant of Java. There is a USAT Virtual Machine based on the Multos virtual machine that runs programs written in C. And there is one based on Microsoft's Smart Card for Windows Runtime Environment that runs programs written in Visual Basic, C, or Java.

There is actually a fourth (and Forth!) virtual machine that is a candidate for use as a USAT Virtual Machine but it hasn't been implemented as such as yet. It is the OTA virtual machine from Europay. This virtual machine has been submitted to ISO for standardization but it has not been made available to the public domain as yet.

The USAT Virtual Machine and SIM Toolkit Programs

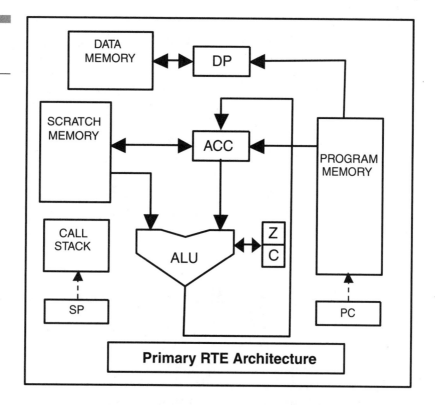

Figure 12-2
Microsoft USAT Virtual Machine.

Unlike the microbrowser, where the two commercial implementations were smashed together to make a standard version, there has been no effort to merge these three virtual machines. As a result, there is no standard-defined set of byte codes for the USAT Virtual Machine as there is for the USAT Interpreter, and each implementation of the USAT Virtual Machine tends to go its own way. You can't move byte codes from one manufacturer's USAT Virtual Machine to another's.

The API to the USAT Virtual Machine has been codified in 3G standards documents and language-specific bindings to this API produced for the Java Card, C, and Visual Basic programming languages. As a result, although applications aren't portable at the byte code level they are portable at the source code level (within each language, of course).

Software development kits for USAT virtual machine programming are surprisingly hard to come by and surprisingly expensive if you can find them. All SIM card manufacturers make SDKs but they sell their kits only to telecom operators. They cost thousands and in some cases tens of thousands of dollars.

Happily there is one kit that is available for free and it supports all three languages. It is the Microsoft Smart Card for Windows kit. It is available for download from http://www.microsoft.com/smartcard. The Microsoft USAT Virtual Machine is called the Runtime Environment (RTE) when it is being used for general-purpose smart-card applications. Because of its easy availability to independent application developers, we'll use the RTE for our discussion and example. All the USAT Virtual Machines work the same way and differ only in the technical details of their byte codes, so that anything we learn about the RTE applies equally to the other virtual machines with a simple change in byte code syntax.

Virtual Machine Architectures

There are two kinds of virtual machines: stack machines and register machines. The Java Card virtual machine is a strict stack machine. The Microsoft RTE is primarily a register machine. This is not a great difference. For example, to add two numbers, A and B, on a stack machine you'd use the following byte codes:

```
PUSH A
PUSH B
ADD
POP  C
```

On a register machine you'd use the following byte codes:

```
LOAD  A
ADD   B
STORE C
```

More important than whether the virtual machine is a stack machine or a register machine is the issue of the complexity of its byte codes.

The fundamental balance that is being struck in designing a virtual machine and its byte code set is between the size of the code that implements the virtual machine and the size of the programs that it interprets. It's CISC versus RISC all over again. The Java Card virtual machine is more like a CISC machine, with a larger virtual machine and smaller application programs. The Microsoft RTE is more like a RISC machine: it has a smaller virtual machine and larger application programs. If you're going to have only a few programs stored on the SIM, then from a storage perspective you're better off with a RISC virtual machine. If you are going to store a lot of programs on the SIM, then you are better off with a CISC virtual machine. The transition (Figure 12-3) from few to many, of course, is the name of the game and takes quantitative measurement that we aren't going to attempt in this brief chapter.

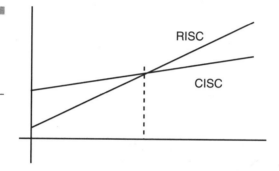

Figure 12-3
Transition from RISC to CISC as more programs are stored on the SIM.

The neat graph shown in Figure 12-3 is tempered a bit by the quality of the compilers that compile the procedural languages into the byte codes and by the density of the byte codes themselves. The RTE has very dense byte codes but, at least initially, not terribly impressive compilers.

There are two other factors that have to be taken into account in smart-card virtual machines: the size of the runtime library and the amount of runtime checking. The runtime library is used by all the applications, so it essentially can be added into the size of the virtual machine itself. Runtime checking is intended to ensure that the application can't do anything it isn't supposed to do and adds to the size of the virtual machine and the slowness of the execution of an application's byte codes. Again, generally speaking, the more complex the byte code set, the more runtime checking has to be done to ensure that the byte codes are behaving themselves. Because all of this checking is based on the hope of heuristics rather than certainty of mathematical proofs, you might find that, the more complex the byte code set, the less certainty that faith in the heuristics is justified.

The USAT Virtual Machine from Microsoft

Figure 12-2 shows a diagram of the Microsoft USAT Virtual Machine. For those with long memories and fading eyesight, it's like an Intel 8048, the Harvard architecture processor that powered one of the first personal computers, the IMSAI.

There are 29 byte codes in the primary RTE virtual machine, and they are listed in Table 12-2. These are called the primary byte codes because as you probably surmised from the ESC byte code, you can bug out to secondary or extended byte codes. One set of extended byte codes, called the *math extension* (Figure 12-4) is provided. Whereas the primary byte codes are based on a register architecture, the math extension byte codes are based on a stack architecture. The 31 math extension byte codes are listed in Table 12-3.

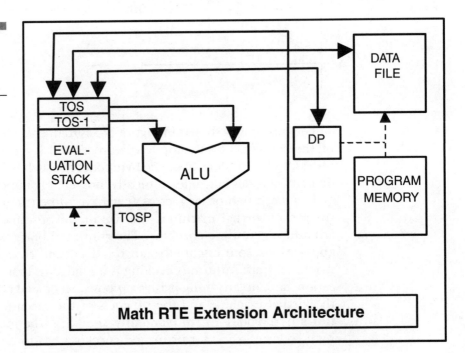

Figure 12-4
Math extension for the Microsoft USAT Interpreter.

The USAT Virtual Machine and SIM Toolkit Programs

TABLE 12-2

USAT Virtual Machine Primary Byte Codes

Mnemonic	Description
ADD	Add a byte to the accumulator
SUB	Subtract a byte from the accumulator
AND	AND a byte with the accumulator
OR	OR a byte with the accumulator
XOR	XOR a byte with the accumulator
LDA, LDAI, LDAD, LDAC	Load a byte into the accumulator
STA, STAI, STAD	Store a byte from the accumulator
JMP	Jump to a location
JZ, JNZ, JC, JNC, JLTZ, JGEZ	Jump based on a condition
CALL	Call a subroutine
RET	Return from a subroutine
CLR	Clear an address
DEC	Decrement an address
INC	Increment an address
NOP	Do nothing
END	Terminate byte code execution
STOP	Debug mode break point
ESC	Transfer control to an extension
SYS	Call a system routine

TABLE 12-3

Arithmetic Extension Byte Codes

Mnemonic	Description
ADD	Add top 2 stack elements and push result
SUB	Subtract top 2 stack elements and push result
MUL	Multiply top 2 stack elements and push result
DIV	Divide top 2 stack elements and push result
MOD	Module of top 2 stack elements and push result
NEG	Negation of top stack element and push result
PUSH	Push a value of a variable on the stack
POP	Pop the top of the stack to a variable
DUP	Duplicate the top of the stack
SWAP	Swap the top 2 elements on the stack
JMP	Jump to a location
JZ, JNZ, JGEZ, JLTZ	Jump based on the top of the stack
CALL	Call a subroutine
RET	Return from a subroutine
NOP	Do nothing
END	Return control to primary byte code set
SHL, SHR	Shift operations
SETDP, SETDPLSB	Set the double precision register (DP) to a value
AND, OR, XOR, NOT	Logical operations on the top of the stack
POPTODP	Pop top of the stack to the double precision register
CMP	Compare values on the top of the stack and set condition flags
PUSHU8, PUSHU16	Push a literal onto the top of the stack

The USAT Virtual Machine and SIM Toolkit Programs

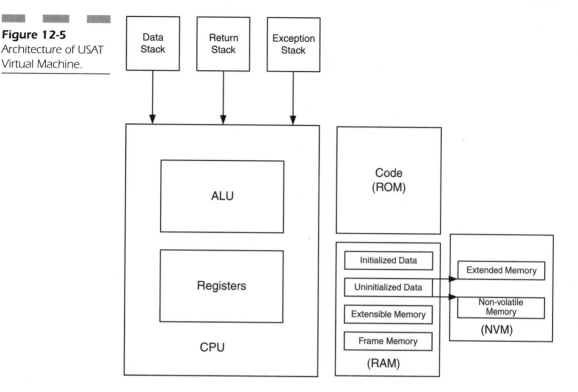

Figure 12-5
Architecture of USAT Virtual Machine.

Another byte code extension could, of course, be the USAT Interpreter byte codes, but this hasn't been done by anybody to our knowledge.

The other wild-card byte code in the primary byte codes set is SYS, the bug out to system routines. This byte code lets your RTE program access the standard Smart Card for Windows API, but it also can access some GSM-specific system services. These GSM-specific system services are listed in Table 12-4.

The Microsoft USAT Virtual Machine supports building applications in Visual Basic and in C. We are died-in-the-wool C hackers, so we'll use C in the following discussion. If you download the package, you'll find a help file that describes the Visual Basic C interfaces.

The GSM runtime library for C programs is listed in Table 12-5.

TABLE 12-4

GSM-Specific System Calls

Load Data Object to Output File
Load Last Data Object and Send Proactive Command
Update GSM Registry
Get Free Timer Number
Release Timer Number
Pop Top of Stack Number to Buffer Indexed
Decode GSM General Result Byte
Parse GSM Terminal Response Bytes
Parse GSM Terminal Response String
Convert Size to GSM Length
End Proactive Command Blocking
Start Proactive Command Blocking
Find Alpha Identifier in SMS-PP File
Send SMS from SMS-PP File

TABLE 12-5

C Runtime Library for the Microsoft USAT Virtual Machine

CatBlockProactiveCommands	CatGetTimer
CatUnblockProactiveCommands	CatGetTimerValue
CatAddItem	CatMoreTime
CatSelectItem	CatPowerOnCard
CatEndSelectItem	CatPowerOffCard
CatEndSelectItemWithIcon	CatPerformCardAPDU
CatDisplayText	CatGetReaderStatus
CatDisplayTextWithIcon	CatExecuteProactive
CatPlayTone	CatProvideLocalInformation
CatSendDTMF	CatGetExtendedError
CatSendDTMFWithIcon	CatGetInKey
CatPollingOff	CatGetInKeyWithIcon
CatPollInterval	CatGetInput
CatRefresh	CatGetInputWithIcon
CatFreeTimer	CatSetupIdleModeText
CatStartTimer	CatRunATCommand

continued on next page

The USAT Virtual Machine and SIM Toolkit Programs

TABLE 12-5

C Runtime Library for the Microsoft USAT Virtual Machine (continued)

CatRunATCommand	CatSendUSSD
CatRunATCommandWithIcon	CatSendUSSDWithIcon
CatSendSMS	CatSetupCall
CatSendSMSWithIcon	CatSetupCallWithIcon
CatSendSMSFromSMSPFile	CatUpdateMenu
CatSendSS	CatUpdatePhysicalEvent
CatSendSSWithIcon	CatUpdateRegistry

Table 12-6 lists some utility functions that let you manipulate the data coming from and going to the handset and build proactive command TLVs byte by byte.

TABLE 12-6

Utility Functions

For use with the GSM Runtime API
   CatPackFromAscii
   CatPackFromUnicode
   CatUnpackToAscii
   CatUnpackToUnicode
Low-level TLV functions
   CatResetBuf
   CatPutByte
   CatLoad
   CatSend
   CatPutLength
   CatParseResponse
   CatParseResponseWithDCS
   CatGetResult
   CatSendAndGetResult
High-level TLV functions
   CatStartCommand
   CatAppendTLV
   CatAppendTLVWithBytePrefix
   CatAppendImmediateResponseTL
   CatAppendIconIdentifierTLV
   CatAppendTextStringTLV(
   CatAppendEncodedTextStringTLV
   CatAppendDurationTLV
   CatAppendAlphaIdentifierTLV

Real-Time Travel Example

Let us recast our stranded vagabond example from Chapter 10 as a USAT Virtual Machine application rather than as a series of USAT Interpreter pages.

Our traveler is sitting in the American Airlines Admiral's Club when an SMS message arrives directly from the Real-Time Travel (RTT) server. The RTT server at RTT Central has been monitoring all the airlines and their flights and has picked up on a developing problem with the traveler's itinerary.

The SMS message contains the notification that the flight the traveler is waiting to board has been delayed. It also includes some alternative flight proposals. The RTT application on the SIM parses the incoming SMS message and displays the situation and the alternatives on the screen. The traveler picks one and the RTT application sends the selection back to RTT Central. In a couple of minutes, a confirmation of the new itinerary arrives back from RTT Central via another SMS message. The RTT USAT Virtual Machine application on the SIM informs the traveler of the arrival of the confirmation and offers a menu to send a notification of the change in travel plans to selected

Central versus Local Storage of Personal Information

The sending of SMS messages to colleagues could be done from the RTT server rather than from the phone, but this would mean storing all the contact information at RTT Central. If you use only one network service, then a strong argument can be made to store semipermanent personal context information such as your address book on a network server.

If you use more than one network service, storing this information at each service means that you have to update each service when it does change. If you store your personal context information locally, on your mobile, and provide it as needed to the services you use, then you only have to update the information in one place when the information changes. The local architecture has the additional advantage that you always know where your information is—it's in your pocket.

business associates and personal contacts. The selection will be made from an address book contained in the SIM.

Because this is a USAT Virtual Machine application and not a USAT Interpreter application, an intermediate network server such as a WIG doesn't have to translate the messages. This mean that messages move faster and that RTT can use the 03.48 standard to achieve end-to-end security between RTT Central and the RTT application on the mobile.

We will assume that the RTT application has been configured on the SIM so that it will be informed of any incoming SMS messages. We will also assume that all 03.48 processing has been successfully performed on the SIM so that our application code gets the message in the clear.

The RTT USAT Virtual Machine application will be looking for messages from RTT Central, of which there are two kinds: notification of an event against an active itinerary and confirmation of a change in an active itinerary. For notification messages, the application code just has to display the message and the options on the screen, get the traveler's selection of one of the options, and return the selection via an SMS message to RTT Central. For confirmation messages, the application code has to forward the message to a predefined list of people whom the traveler wants to alert to changes in his travel plans.

```
/*
** Real-Time Travel SIM Toolkit Application
*/

#include <string.h>
#include <gsm.h>

#define NOTIFICATION 0x01
#define CONFIRMATION 0x02

/* Raw Incoming SMS Message */
static BYTE SMS_Deliver[180];

/* Telephone number of SMS message center at RTT Central */
static RTTOriginatingAddress[] = {0x61, 0x71, 0x32, 0x10, 0x43, 0xF6};

/* Reply template for chosen notification option */
char *rttNotificationReply = "X - Customer 1234-56";
```

```c
/* Number of people to be notified if travel plans change */
BYTE rttCoordinators = 5;

/* Status Words */
BYTE SW_OK[] = {0x90, 0x00};
BYTE SW_NO[] = {0x6A, 0x00};

void main(BYTE cla, BYTE ins, BYTE P1, BYTE P2, TCOUNT Lc)
{
  BYTE *rttMsg, rttItems, *rttMsgp, rttOptions, rttChoice;
  int i;

  /*
  ** Check to see if message is from RTT Central. The
  ** message is an SMS-DELIVER in the data field of the
  ** ENVELOPE APDU.
  */
  ScwGetCommBytes(SMS_Deliver, Lc);

  /* Make sure the SMS is from RTT Central */
  if(SMS_Deliver[1] != 0x0B || SMS_Deliver[2] != 0x91 ||
     memcmp(&SMS_Deliver[3], RTTOriginatingAddress, 6) != 0) {
    ScwSendCommBytes(SW_NO, sizeof(SW_NO));
       return;
  }

  rttMsg = &SMS_Deliver[19];

  /* Process the message from RTT Central */
  switch(rttMsg[0]) {

  case NOTIFICATION:
    /*
    ** Put the notification and list of options on
    ** the screen and get a choice
    */
    rttItems = rttMsg[1];
    GsmSelectItem(rttMsg[2], &rttMsg[3],
                       DEFAULT_STYLE_NO_HELP);
```

The USAT Virtual Machine and SIM Toolkit Programs

```
      rttMsgp = &rttMsg[rttMsg[2]+2];
      for(i = 0; i < rttOptions; i++) {
        GsmAddItem(*rttMsgp, rttMsgp+1, i+1);
        rttMsgp += *rttMsgp+1;
      }
      GsmEndSelectItem(&rttChoice);

      /* Send the choice back to RTT Central */
      rttNotificationReply[0] = rttChoice+0x30;
      GsmSendSMSFromSMSPFile(sizeof(rttNotificationReply),
                   (BYTE *)rttNotificationReply, 0);

      ScwSendCommBytes(SW_OK, sizeof(SW_NO));

      break;

    case CONFIRMATION:

      /* Display the confirming message from RTT Central */
      GsmDisplayText(DCS_SMS_UNPACKED, rttMsg[0], &rttMsg[1],
                              HIGH_PRIORITY_USER_CLEAR, 1);

      /* Forward the mesage to each travel coordinators */
      for (i = 0; i < rttCoordinators; i++)
        GsmSendSMSFromSMSPFile(rttMsg[0],&rttMsg[1], i+1);

      ScwSendCommBytes(SW_OK, sizeof(SW_OK));

      break;

    default:
      ScwSendCommBytes(SW_NO, sizeof(SW_OK));
      break;
  }
}
```

This compiles to the 387 bytes of executable USAT Virtual Machine byte codes. So you can see what a complete byte code rendering of a C program looks like, we list the complete compilation.

```
SECTION .text

    000007                  .text$$start
    000007                  start
    000007    2400              JMP _main
    000009                  _main
    000009    be0100            SETDP @0
    00000c    78                MATHRTE
    00000d    48                SETBYTEMODE
    00000e    60                POP 0
    00000f    0d0101            SETDP @1
    000012    60                POP 0
    000013    0d0102            SETDP @2
    000016    60                POP 0
    000017    0d0103            SETDP @3
    00001a    60                POP 0
    00001b    0d0104            SETDP @4
    00001e    60                POP 0
    00001f    4b                SETUNSIGNEDMODE
    000020    70                PUSH 0
    000021    0240              PUSH #_SMS_Deliver
    000023    44                PRIMARYRTE
    000024    21f7              CALL _ScwGetCommBytes
    000026    be0041            SETDP _SMS_Deliver + 1
    000029    3741              SYS RDDTFILE
    00002b    3a0b              SUB #11
    00002d    1418              JNZ @5
    00002f    be0042            SETDP _SMS_Deliver + 2
    000032    3741              SYS RDDTFILE
    000034    3a91              SUB #145
    000036    140f              JNZ @6
    000038    7c06              PUSH #6
    00003a    7c0c              PUSH #_RTTOriginatingAddress
    00003c    7c43              PUSH #_SMS_Deliver + 3
    00003e    2145              CALL _memcmp
    000040    78                MATHRTE
    000041    f2c3              NOP JNZ @7
    000043    44                PRIMARYRTE
    000044    2408              JMP @8
    000046                  @7
    000046    44                PRIMARYRTE
```

The USAT Virtual Machine and SIM Toolkit Programs

```
000047                    @6
000047                    @5
000047    7c02            PUSH #2
000049    7c29            PUSH #_SW_NO
00004b    21bf            CALL _ScwSendCommBytes
00004d    fe              EXIT
00004e                    @8
00004e    7c53            PUSH #_SMS_Deliver + 19
000050    be00f4          SETDP @9
000053    78              MATHRTE
000054    49              SETINTMODE
000055    60              POP 0
000056    70              PUSH 0
000057    4f              POPTODP
000058    48              SETBYTEMODE
000059    4b              SETUNSIGNEDMODE
00005a    70              PUSH 0
00005b    44              PRIMARYRTE
00005c    3761            SYS GETBYTE1FROMTOS
00005e    78              MATHRTE
00005f    40              POPONE
000060    44              PRIMARYRTE
000061    d1              SUB #1
000062    1005            JZ @10
000064    d1              SUB #1
000065    10b2            JZ @11
000067    2512            JMP @12
000069                    @10
000069    be00f4          SETDP @9
00006c    78              MATHRTE
00006d    49              SETINTMODE
00006e    70              PUSH 0
00006f    4f              POPTODP
000070    48              SETBYTEMODE
000071    71              PUSH 1
000072    0d00f6          SETDP @13
000075    60              POP 0
000076    50              PUSH #0
000077    0d00f4          SETDP @9
00007a    49              SETINTMODE
00007b    70              PUSH 0
```

00007c	53		PUSH #3
00007d	f9		NOP ADD
00007e	70		PUSH 0
00007f	4f		POPTODP
000080	48		SETBYTEMODE
000081	72		PUSH 2
000082	44		RIMARYRTE
000083	21ff		CALL _GsmSelectItem
000085	be00f4		SETDP @9
000088	78		MATHRTE
000089	49		SETINTMODE
00008a	70		PUSH 0
00008b	4f		POPTODP
00008c	48		SETBYTEMODE
00008d	72		PUSH 2
00008e	52		PUSH #2
00008f	f9		NOP ADD
000090	0d00f4		SETDP @9
000093	49		SETINTMODE
000094	70		PUSH 0
000095	f9		NOP ADD
000096	0d00f8		SETDP @14
000099	60		POP 0
00009a	50		PUSH #0
00009b	0d00fc		SETDP @15
00009e	4a		SETLONGMODE
00009f	60		POP 0
0000a0	44		PRIMARYRTE
0000a1	2437		JMP @16
0000a3		@17	
0000a3	be00ff		SETDP @15 + 3
0000a6	78		MATHRTE
0000a7	48		SETBYTEMODE
0000a8	4b		SETUNSIGNEDMODE
0000a9	70		PUSH 0
0000aa	51		PUSH #1
0000ab	f9		NOP ADD
0000ac	0d00f8		SETDP @14
0000af	49		SETINTMODE
0000b0	70		PUSH 0
0000b1	51		PUSH #1

The USAT Virtual Machine and SIM Toolkit Programs

```
0000b2    f9                NOP ADD
0000b3    70                PUSH 0
0000b4    4f                POPTODP
0000b5    48                SETBYTEMODE
0000b6    4b                SETUNSIGNEDMODE
0000b7    70                PUSH 0
0000b8    44                PRIMARYRTE
0000b9    21d6              CALL _GsmAddItem
0000bb    be00f8            SETDP @14
0000be    78                MATHRTE
0000bf    49                SETINTMODE
0000c0    70                PUSH 0
0000c1    4f                POPTODP
0000c2    48                SETBYTEMODE
0000c3    4b                SETUNSIGNEDMODE
0000c4    70                PUSH 0
0000c5    51                PUSH #1
0000c6    f9                NOP ADD
0000c7    0d00f8            SETDP @14
0000ca    49                SETINTMODE
0000cb    70                PUSH 0
0000cc    f9                NOP ADD
0000cd    60                POP 0
0000ce    44                PRIMARYRTE
0000cf              @18
0000cf    7c01              PUSH #1
0000d1    be00fc            SETDP @15
0000d4    78                MATHRTE
0000d5    4a                SETLONGMODE
0000d6    70                PUSH 0
0000d7    f9                NOP ADD
0000d8    60                POP 0
0000d9    44                PRIMARYRTE
0000da              @16
0000da    be00fa            SETDP @19
0000dd    78                MATHRTE
0000de    48                SETBYTEMODE
0000df    4b                SETUNSIGNEDMODE
0000e0    70                PUSH 0
0000e1    0d00fc            SETDP @15
0000e4    4a                SETLONGMODE
```

Chapter 12

0000e5	70		PUSH 0
0000e6	fd		NOP CMP
0000e7	f203		NOP JLEZ @20
0000e9	44		PRIMARYRTE
0000ea	27b7		JMP @17
0000ec		@20	
0000ec	44		PRIMARYRTE
0000ed		@21	
0000ed	7dfb00		PUSH #@22
0000f0	21a6		CALL _GsmEndSelectItem
0000f2	78		MATHRTE
0000f3	40		POPONE
0000f4	0d00fb		SETDP @22
0000f7	48		SETBYTEMODE
0000f8	4b		SETUNSIGNEDMODE
0000f9	70		PUSH 0
0000fa	0230		PUSH #48
0000fc	f9		NOP ADD
0000fd	0d0024		SETDP _rttNotificationReply
000100	49		SETINTMODE
000101	70		PUSH 0
000102	4f		POPTODP
000103	48		SETBYTEMODE
000104	60		POP 0
000105	50		PUSH #0
000106	0d0024		SETDP _rttNotificationReply
000109	49		SETINTMODE
00010a	70		PUSH 0
00010b	52		PUSH #2
00010c	44		PRIMARYRTE
00010d	2075		CALL _GsmSendSMSFromSMSPFile
00010f	78		MATHRTE
000110	40		POPONE
000111	52		PUSH #2
000112	0227		PUSH #_SW_OK
000114	44		PRIMARYRTE
000115	20f5		CALL _ScwSendCommBytes
000117	246a		JMP @23
000119		@11	
000119	7c01		PUSH #1

The USAT Virtual Machine and SIM Toolkit Programs

```
00011b   7d8100         PUSH #129
00011e   be00f4         SETDP @9
000121   78             MATHRTE
000122   49             SETINTMODE
000123   70             PUSH 0
000124   51             PUSH #1
000125   f9             NOP ADD
000126   70             PUSH 0
000127   4f             POPTODP
000128   48             SETBYTEMODE
000129   4b             SETUNSIGNEDMODE
00012a   70             PUSH 0
00012b   54             PUSH #4
00012c   44             PRIMARYRTE
00012d   2175           CALL _GsmDisplayText
00012f   78             MATHRTE
000130   40             POPONE
000131   50             PUSH #0
000132   0d00fc         SETDP @15
000135   4a             SETLONGMODE
000136   60             POP 0
000137   44             PRIMARYRTE
000138   2426           JMP @24
00013a              @25
00013a   be00ff         SETDP @15 + 3
00013d   78             MATHRTE
00013e   48             SETBYTEMODE
00013f   4b             SETUNSIGNEDMODE
000140   70             PUSH 0
000141   51             PUSH #1
000142   f9             NOP ADD
000143   0d00f4         SETDP @9
000146   49             SETINTMODE
000147   70             PUSH 0
000148   51             PUSH #1
000149   f9             NOP ADD
00014a   70             PUSH 0
00014b   4f             POPTODP
00014c   48             SETBYTEMODE
00014d   4b             SETUNSIGNEDMODE
00014e   70             PUSH 0
```

```
00014f   44                    PRIMARYRTE
000150   2032                  CALL
                               _GsmSendSMSFromSMSPFile
000152   78                    MATHRTE
000153   40                    POPONE
000154   44                    PRIMARYRTE
000155              @26
000155   7c01                  PUSH #1
000157   be00fc                SETDP @15
00015a   78                    MATHRTE
00015b   4a                    SETLONGMODE
00015c   70                    PUSH 0
00015d   f9                    NOP ADD
00015e   60                    POP 0
00015f   44                    PRIMARYRTE
000160              @24
000160   be0026                SETDP _rttCoordinators
000163   78                    MATHRTE
000164   48                    SETBYTEMODE
000165   4b                    SETUNSIGNEDMODE
000166   70                    PUSH 0
000167   0d00fc                SETDP @15
00016a   4a                    SETLONGMODE
00016b   70                    PUSH 0
00016c   fd                    NOP CMP
00016d   f203                  NOP JLEZ @27
00016f   44                    PRIMARYRTE
000170   27c8                  JMP @25
000172              @27
000172   44                    PRIMARYRTE
000173              @28
000173   7c02                  PUSH #2
000175   7c27                  PUSH #_SW_OK
000177   2093                  CALL _ScwSendCommBytes
000179   2408                  JMP @23
00017b              @12
00017b   7c02                  PUSH #2
00017d   7c29                  PUSH #_SW_NO
00017f   208b                  CALL _ScwSendCommBytes
000181   2400                  JMP @23
000183              @23
000183   fe                    EXIT
```

Java Card™ SIMs

The most advanced USAT Virtual Machine in the standardization process is the Java Card™ SIM. A Java Card™ SIM is created by simply putting a Java Card™ virtual machine onto a SIM card and providing it with runtime library access to the SAT API.

Unfortunately, there are a number of unresolved intellectual property rights questions with regard to the use of the Java Card™ SIM. The Java Card™ SIM is described at length in 3GPP TS 03.19 SIM API for Java Card™ and the reader is referred there for details (Figure 12-6).

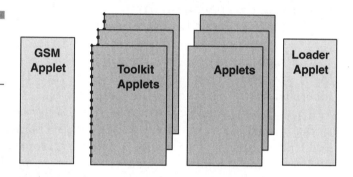

Figure 12-6
Architecture of the Java Card™ USAT Virtual Machine.

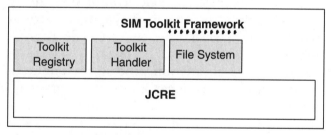

•••••• shareable interface

Installation of USAT Virtual Machine Programs

In one way or another, each USAT Virtual Machine implementation writes the byte codes of an application program to a file on the SIM as the first step in preparing the program for execution. It is this step that differentiates USAT Virtual Machine applications from USAT Interpreter applications and is the most significant difference in the administration and maintenance expense between these two ways of building a mobile application.

Clearly, the network operator can't allow just anybody (including the subscriber) to fill up the precious EEPROM of the SIM with application byte codes. As a result, there is a serious, industrial-strength security protocol that must be followed to actually get the application onto the SIM, whether at a kiosk or over the air. A number of commercial protocols are available to handle this phase of application installation. The two most widely used are the Global (formerly Visa Open) Platform protocol and the Multos Load Certificate protocol.

After the byte codes of the application have been written to a file on the SIM, the application has to be registered so that the SIM knows it's there and available for execution. Otherwise, the SIM would think that the byte code file contained some undifferentiated random data. In the registration process, the application's application identifier (AID) is written into a file called EF_{DIR} in the MF, which has the file identifier 0x2F00, for those who really care.

EF_{DIR} also contains a human-readable description of the application so that the handset can display a list of all the applications on the SIM and let the subscriber start one by picking it from this list. Because the SIM operating system has associated each entry in EF_{DIR} with the file that contains the byte codes comprising the application, when an entry is picked, the operating system can activate the application by handing the file of byte codes to the appropriate USAT Virtual Machine.

During the installation process, the application is also given the opportunity to say what kinds of events it is interested in being informed about. In our RTT example, our application asked to be informed of all incoming SMS-PP messages during this phase of installation.

That's it. Write the application. Compile it to byte codes. Download it to the SIM. Install it. And you're ready to go. Creating USAT Virtual Machine applications is all very straightforward from technical point of view and not a bit different that what you do with applications on a mainframe or PC. The hard part is convincing a network operator to let you install your application on the SIM and this is what makes the USAT Virtual Machine rendering of mobile applications so much more difficult than using the USAT Interpreter rendering of mobile applications.

Summary

The USAT Interpreter and the USAT Virtual Machine provide standardized high-level language access to the SAT API. The significant difference is that USAT Interpreter programs are not permanently installed on the SIM and USAT Virtual Machine programs are.

The USAT Interpreter approach to mobile applications is so far ahead of the USAT Virtual Machine approach because it provides more control to the network operator and significantly reduces administrative costs.

It makes no sense for a SIM to carry two byte code interpreters, one for permanent programs and one for transient programs. A unified interpreter that can handle both types of programs is technically feasible. Due to the rapid take up of the USAT Interpreter mode of application development, this unified interpreter probably will evolve from the USAT Interpreter.

CHAPTER **13**

Smart Signatures for Secure Mobile Commerce

Fitting the greatest possible functionality into the smallest number of keystrokes is a major challenge for wireless application developers. Even as mobile devices and bandwidth speeds become more attuned to transaction and data-centric applications, the drive to streamline applications and enhance functionality won't let up. Balancing fast and simple interfaces with requirements for security and flexibility always will be a critical element in successful m-commerce offers. And users on the go will gravitate toward the convenience of services that don't require a lot of effort to get up and running. This was the thinking behind the design of Smart Signature[1], an application implemented by SmartTrust to combine the convenience of customized menus and templates for various m-commerce services with the security of PKI and digital signatures.

One of the guiding principles of the SmartSignature solution was that it should only take one SMS message for the user to activate a m-commerce transaction, complete with digital signature. Imposing this strict limit made development a challenge, but it also had a number of benefits. From the network and mobile service provider's point of view, the application would be bandwidth efficient by delivering rapid response time even at 2G connection speeds. The customer could take advantage of multiple services at a very low cost and with minimal effort. Instead of sending a string of text messages to complete every transaction, the user could take advantage of the built-in features of a menu-driven application.

How much can a mobile phone user really accomplish by pushing a couple of keys and sending just one SMS message? A surprising amount, when that message is driven by a well-designed SIM Toolkit application with a built-in SAT Interpreter and menus specifically configured to the task at hand. SmartSignature is one of the few working GSM2 applications that enables the exchange of digitally signed and confidential short messages using public key technology. From a customer perspective, SmartSignature packs a lot of value into a small package:

1. Sending and receiving encrypted short messages that can only be deciphered at the service provider's premises (i.e., its dedicated server) or in the customer's mobile phone. This ensures confidentiality between a customer and a service provider (SP).

---

[1] SmartSignature is a product that has been developed by SmartTrust with patented technology.

2. Sending and receiving of digitally signed short messages. A certified public key verifies the signature. This is the first step in ensuring message integrity and providing nonrepudiation guarantees.
3. A clear separation between digital keys dedicated for signing messages and keys used for encrypting and decrypting messages.
4. Usage of public key certificates that are based on standard X.509 certificates and issued by existing commercial certification authorities. A trusted third party that is not responsible for the mobile network infrastructure or the mobile service provisioning handles the chain of trust and the certification of the public keys. This entails the use of standardized "Internet-based" PKI components and certification services.

Delivering a PKI-based security solution that takes full advantage of SIM and SMS capabilities is an impressive feat, but as this SmartSignature case will show, the business and trust relationships required to create a working PKI infrastructure for a specific mobile operator and service provider are even more complex then the technical development issues. As many application developers will discover, the ultimate success of a SIM or SMS application will depend as much on its commercial viability as on its technical merits. By implementing a large-scale delivery pilot for Smart Signature, SmartTrust was able to analyze the critical business success factors for a future rollout of this solution.

Starting With the Mobile Customer

Before exploring the extended m-commerce value chain required for a large-scale mobile PKI implementation, let's zero in on SmartSignature from the mobile user's point of view and see how the application design decisions made by SmartTrust help to enhance that user's experience.

Imagine the following scenario, which is enabled by the Smart-Signature application:

After selecting a banking service provider that offers SmartSignature, Christine decides to do away with the hassle of writing paper checks and signs up to pay her car loan installments directly from her mobile phone. The bank sends her a payment order in a short message, which is signed with the bank's digital key and encrypted specifically for Christine. This way Christine can verify that the sender is really the bank and that the content has not been tampered with along the way.

Signed and encrypted short messages are a particular type of SmartSignature message, and they are stored in a dedicated SmartSignature area on Christine's SIM that is different from the conventional short message folder. After such a message arrives, Christine is notified of its arrival. With a couple of menu picks, she can decrypt the message with her own digital key and verify the message with the bank's digital signature. To reply, Christine can simply digitally sign the order to authorize the payment or she can call up a complementary template that lets her change the amount of the payment if she wants to pay her loan off more quickly.

This has been a good month for overtime work, so she picks the "modify" option to add an extra $50 to the payment total. Just a few keystrokes are needed to change the payment amount, and then Christine clicks on the "Sign" button. Christine enters her *signing PIN* for authorizing the signing process that relies on her private signing key. After signing the payment order, Christine enters the send command with a single keystroke. Before sending back the authorized payment order, Christine has the option* to review the entire signed SMS message to double check that everything is just the way she wants it. Once Christine has chosen to send the signed message, the message will be encrypted with the bank's public encryption key. As a result, only the receiving bank can decrypt Christine's message.

Another reason Christine could afford the extra car payment this month is that she was able to take advantage of a special holiday airline fare offered by her favorite airline. This reduced fare was available only to frequent fliers using the SmartSignature-based discount travel alert program. Christine confirmed her ticket reservations on her mobile phone with the use of a Smart-

> Signature menu similar to that of the banking service but customized for verifying and accepting travel arrangements. Knowing that the entire transaction was secure and authenticated with digital signatures from the airline and her travel agent meant that she didn't have to worry about the security of her transaction or the validity of her electronic ticket confirmation.
>
> ---
>
> *Some national legislation for digital signatures (for example, in Germany) require that a customer sees the entire signed message content before it is sent to the SP. In this respect, the viewing of the signed content is not an option but a feature mandated by legislation. The requirements for this particular feature are based on the "what you see is what you sign" constraint. That also means that the signed short message does not contain any "hidden" data such as tracking numbers.

SmartSignature Features

Forms and Templates

This scenario illustrates some of the key features of the SmartSignature design. First and foremost of these are the application-specific forms (i.e., templates) that are an integral part of situation-specific transaction services. Not only do the carefully designed templates allow the subscriber to quickly and efficiently interact with the application server residing at the service provider, but also they are a part of the engineering concept that enables the "single-SMS-message" design goal of the overall system.

SmartSignature messages are based on the concept of a *Form*. Each SmartSignature message consists of two parts: a static template, known as *FormPage*, and dynamic data, known as *FormData*. The static template is stored in the SmartSignature section of the SIM and the dynamic data arrives as an SMS message. When the dynamic data is combined with the static template, the result is a Form, which is shown to the user as a message. This combination of storing the static template data on the SIM and sending only the transaction-specific dynamic data enables the one-message design goal of the SmartSignature system to become a reality. This simplicity doesn't reduce security, however, because the FormPage and FormData are transferred to the customer's mobile phone in the secure delivery mode (using signed and encrypted short messages).

Depending on the template settings, the user manipulates the incoming dynamic data in several ways. A typical FormPage includes text input, list selection, default values, and display and approve options. For example, with a bill presentment and payment transaction, the user will open an incoming message, display the amount due, decide whether to pay that amount or modify it, and then approve the final transaction.

Keys and PINs

The security on the SIM includes two RSA private keys owned by the user. One private key is used for signing messages that are sent to SPs. The other private key is used for decrypting messages that have been sent by service providers. Service providers use the user's public encryption key, available from a (digital) X.509 certificate in the directory, for encrypting messages before their transport over the fixed and wireless network to the mobile phone user. The certificate directory is a repository for digital certificates issued for a specific RSA public/private key pair. The public key can be found in the certificate itself, whereas the party who has been certified knows only the corresponding private key.

The private keys on the SIM card for the mobile phone user are stored in the tamper-proof area. The mobile phone user who intends to sign a message has to activate the signing process by submitting a *signing PIN*. The user's private key is never exposed to any third party; it just engages the algorithm that performs the signing function in SmartSignature. The signing PIN simply authorizes the algorithm to use the private key. This implementation is a standard procedure in marrying digital signature implementations with the legal requirements for keeping private keys in a secure, exclusive-access "area."

Menu Design

When subscribers (like Christine) have signed up for airline travel and banking services from two separate providers, they see a menu that is organized as shown in Figure 13-1.

At the top-level menu, where the names of the service providers "bank," "airline," etc., are displayed, an asterisk next to a provider's name alerts the user that there are new messages from this provider waiting to be read.

Smart Signatures for Secure Mobile Commerce

Figure 13-1
Top-level menu organization.

```
*Bank Name
    Messages [2]
    Fill-in Forms
    Service Options
*Airline Name
    Messages [4]
    Fill-in Forms
    Service Options
```

Each service provider uses the same menu structure with the same three items. The Message heading holds received messages and tells the user how many new messages are waiting. The Fill-in Forms folder holds FormPages (i.e., templates) that can be removed and added dynamically OTA to suit the needs of the service provider and Service Options opens to a list of service provider-specific options.

The menu structure and items have been designed for a uniform user experience and fast access. However, service providers can use FormPages to differentiate and tailor their services.

A sample screenshot that would be found two levels down the menu structure for a bill payment service is shown in Figure 13-2.

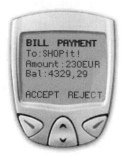

Figure 13-2
Sample bill payment menu (courtesy of Setec at www.setec.fi).

Changing Service Providers

Suppose, however, that the customer wants to change banks or add an entirely new service provider such as a stockbroker or a florist? SmartSignature makes this type of change very easy from the customer's point of view. Each service provider has a code or name, which the user simply enters into her phone. Because there is a separate folder on the SIM for each service provider, the user can add a new service provider and receive a new certificate for that service provider and store the keys that will be used to interact with that service provider with a few keystrokes. As new services are added, the menu displays on the phone will be updated automatically with the new service provider menu item.

The change procedure can be separated into the following discrete steps:

1. The customer can request a new service provider upload by sending a short message with the service provider's codename.
2. Alternatively, the service provider may be authorized to push an upload request to the mobile phone and the user then can decide to accept the uploading request.
3. The mobile operator delivers the service provider's mobile certificate to the phone. The communication is secured and vouched for by the mobile operator.
4. The mobile phone user can verify the service provider's mobile certificate.
5. Then the service provider uploads a set of FormPages.

After step 5, the service provider and mobile phone user are set up for exchanging secure transaction requests and authorizations. This entire set of interactions can take place in just a few minutes and with just a few menu picks on the part of the mobile user.

Christine, for example, can start using her SmartSignature-enabled mobile phone with an airline service and then add on a banking service with its own separate menu. Each new service provider that has signed up to the SmartSignature scheme can ask its target customers to upload their new menu item. In some cases, a service initialization message is pushed to the phone, but then it is always up to the mobile phone user to accept an upload of the service and the service provider's keys.

When the initial message from a new service provider is sent to the user, SmartSignature will immediately notify the user with a one-time secure notification that is displayed on the phone. The mobile operator guarantees that the first message from the service provider is a bona fide message. The customer can tell from the format of this message display that it is coming from a secured service provider and not just a friend who wants to chat. In essence, the whole treatment of incoming SmartSignature traffic is handled differently to enhance security. As previously indicated, the transactions are not stored on the normal SMS side of the SIM; they go into a separate SmartSignature region and the user has to enter a separate Login PIN to access them.

The actual screen display depends on the FormPages used by the service provider. Each service provider can use its own FormPage designs to provide a specific service function to the user. These service provider FormPages are signed by the service provider. The service provider's public keys are certified by a certificate authority, and the mobile operator vouches for the correct transfer of the keys from the certificate directory to the user's mobile phone. Once the FormPage is received, the Smart Signature application verifies the FormPage with that service provider's keys and the customer is ready to do business. Once new services have been activated, the user can check the inbox listing for received messages for all of the selected service providers.

Of course, customers don't want to use certain types of services or service providers forever. Sometimes they want to switch banks or they decide that it isn't that important to order flowers using their phone after all. In these cases, the user can easily remove the unwanted service provider templates from the phone menu by using a built-in delete option. In fact, the customer has several options in changing or deleting service provider functions. The customer can delete a SP totally and all messages will be discarded in the future or can leave the menu intact but request that service provider not send any push messages.

The service providers also have some options on determining how the messages that they send can be handled by the user. A message can be set to be deleted automatically after the user has seen it, e.g., a stock price notification, or other type of share information will be deleted once the user has displayed it. Typically, the most time-sensitive information will be deleted after reading, whereas a bill can be displayed, stored, and checked later for payment; the bill will be deleted when it has been accepted and paid. Because of the limited storage capacity of the SIM, there is no "history" file presently available for past transactions.

The number of service providers that SmartSignature can handle is based on how much memory is on the card and how much space the mobile operator requires for its dedicated network section.

Each service provider requires some amount of space in addition to the various FormPage templates and the messages. The system can instantly accept messages without caching those in memory or storing them.

There is a common space for messages, perhaps 20, from all providers. Once that common area is full, then no more messages can be received. The overflow is stored on the network and the operator will send new messages through once the user clears the memory space.

The benefit of all this from the mobile user's point of view is that the ability to add new services and establish new menu items directly on the wireless phone comes with the security of having every new service provider authenticated through a digital certificate. More detail about the sequence of this process is provided in the next section on the interrelationship of the different entities involved in certifying mobile trust.

Mobile Certification and Trust Using SmartSignature

To understand the thinking behind the PKI implementation by SmartSignature, it is helpful to consider the following questions:

- What kind of certified entities exist in the SmartSignature setting and how are they certified?
- Who is trusting whom during the service setup phase and during a transaction?
- How does SmartSignature obtain the relevant information for establishing a secure transaction between the mobile phone user and a service provider?

Each service provider (or rather the provider's server that runs SmartSignature) is identified as an entity in the Mobile Network Space. Likewise, each SmartSignature implementation on a SIM card constitutes a unique identity. The Network Identity (NID) indicates that the SmartTrust's PKI implementation uses this unique identifier to distinguish the different entities.

The Card Domain (CD) and Root Domain (RD) are two identities allocated to the server that provide certain SmartSignature management services to each phone. In most cases the mobile operator runs and maintains the servers that hold the CD and RD identities (Figure 13-3).

Smart Signatures for Secure Mobile Commerce

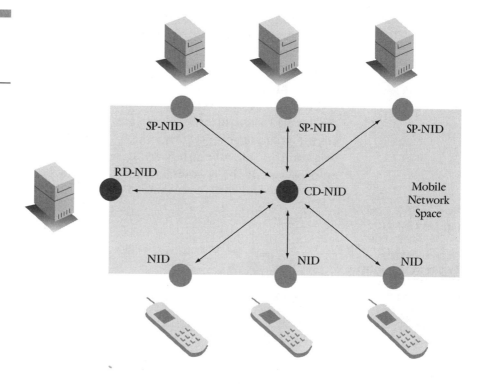

Figure 13-3
Mobile network entities related to providing trust.

Apart from the RD, these entities hold an encryption and signing key pair. The public keys of the key pair are certified and stored in an X.509 certificate.

Space restrictions on the SIM card do not allow SmartSignature to generate the user's key pairs after she has received her SIM card. Therefore, the SIM card vendor preinstalls the key pairs and issues an anonymous X.509 certificate (a.k.a., NID Certificate). Once the mobile phone user has registered herself and received a SmartSignature SIM card, an ID certificate can be issued. The ID certificate contains the same information as the NID certificate plus any information that links a natural person (i.e., the mobile phone user) to the public key. Once an ID certificate has been issued, NID certificates function only as a guarantee from the card vendor that a particular key pair had been preinstalled in SmartSignature during the SIM production process.

Because the storage space on SmartSignature is restricted, Smart-Trust has developed the Mcert (mobile certificate concept). A mobile certificate is an *n*-tuple of the X.509 certificate. In other words, it is an X.509 certificate in condensed form. In the default scenario, the RD and CD Mcerts are also uploaded into SmartSignature during the SIM production process.

The role of the CD is to provide SmartSignature with a secure way to upload new service provider Mcerts and ensure that SmartSignature receives an answer with any certificate status-checking requests for a specific service provider.

The RD facilitates the certificate rollover process for the CD. SmartSignature accepts only control messages that change the CD settings on the SIM when they are signed by the RD. Figure 13-4 provides an overview of the entities that have been issued X.509 certificates.

Figure 13-4
Entities with X.509 certificates in the mobile trust process.

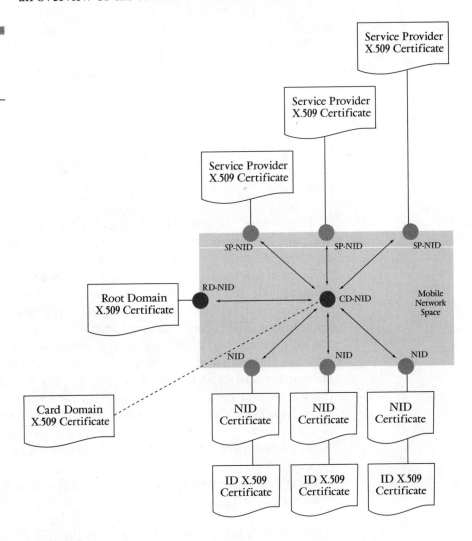

Trust Relationships for Making the Transaction

Once an end user has subscribed to a service from the service provider and has uploaded the service provider's Mcert to the SmartSignature, both parties are ready to communicate using secure transactions. From this point forward, for each transaction the trust is established between the service provider and the mobile phone user. Figure 13-5 illustrates this flow of trust creation for transactions.

Figure 13-5
Trust relationships in place for enabling the transaction.

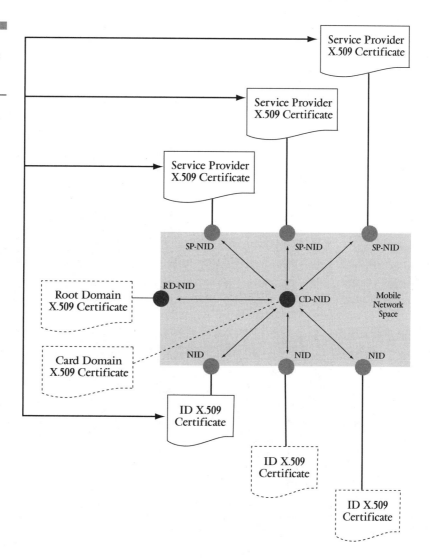

Trust Relationship for Enabling the Transaction

However, for the initial service request to place an service provider Mcert into SmartSignature, the service provider and the mobile phone user have to trust the CD (and RD) certificate, as illustrated in Figure 13-6.

Figure 13-6
Initial creation of required trust relationships.

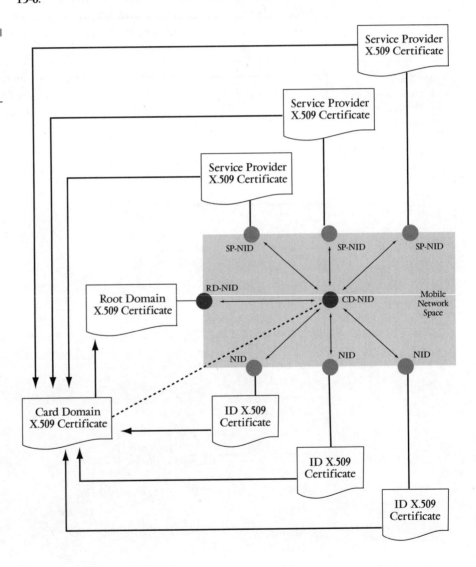

Smart Signatures for Secure Mobile Commerce

Certification Authorities

Having looked at the implementation of the mobile PKI in the SmartSignature setting, it is useful to take a look at the certificate hierarchy and the hierarchy of the certification authorities (CAs) that issue the X.509 certificates. Figure 13-7 depicts one possible scenario.

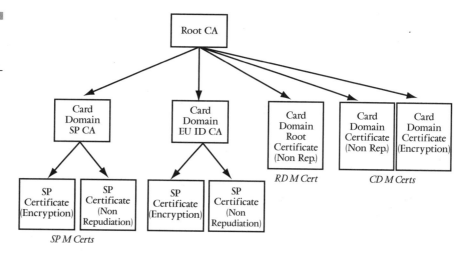

Figure 13-7
Hierarchy of certificate authorities.

In this case there are two CAs that issue certificates for service providers or for individual mobile phone users. The two CAs must have a valid trust relationship, which is indicated by their link to the root CA. The root CA issues CD and RD certificates or it has a trust relationship with a CA that issues RDs. In general, the "certification path" or "chain of trust" is established through CAs that have been providing certification services in the Internet setting.

Business Enablers of SmartSignature

Delivering all this functionality and convenience to the m-commerce customer and the service providers requires more than advanced technology and careful design. It also involves a complex set of agreements and shared responsibilities among multiple companies in the m-commerce value chain. Before examining in greater detail a large-scale pilot delivery of SmartSignature, it is important to appreciate the multitude of business interrelationships that arise during the development setup and operation of a fully functioning SmartSignature solution.

To understand the business context of a product such as SmartSignature, which requires a complex technical infrastructure and a supporting PKI, it is useful to review in more detail the characteristics of an operational system.

A typical communication flow among all the different components needed to enable a SmartSignature operation is illustrated in Figure 13-8. Some of these components, such as the SmartConnection and SmartSecurity servers, are also SmartTrust products.

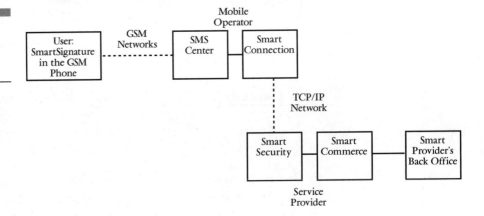

Figure 13-8 Components for SmartSignature in operation.

The next two sections provide more information about the setup and operation requirements of an end-to-end mobile trust infrastructure for SmartSignature. It describes the specific roles of various participants in the mobile trust hierarchy, including the GSM network operators, SIM vendors, and the service providers.

SmartSignature in Operation

Once a mobile phone user exchanges his SIM card for a card that features SmartSignature and the service provider installs SmartCommerce in his application provisioning platform, both have to request

X.509 certificates for their specific RSA public keys that are used to verify signatures or encrypt messages.

The SmartSignature user has access to a set of preinstalled private RSA keys, whereas the service provider has to generate its own RSA key pair and store it in a hardware security module that can be accessed by its SmartTrust system component (i.e., SmartSecurity).

In a simple case, as discussed in the previous section, each certificate is issued by a single CA. Hence, there is a trusted direct link, because only one entity vouches for the issued certificates and the implied link of the respective public key with a legal entity of the service provider and a SmartSignature user.

It is also possible to upload a service provider's credentials (FormPages and Mcert) during the card production process. In such a case, the service provider and the customer are ready for mobile transactions. Both parties can immediately send and receive signed and encrypted short messages.

In general, this is the basic scenario that enables a mobile phone user and service provider to execute confidential transactions. A valid digital signature also guarantees that the message has not been tampered with during transportation. Additionally, depending upon the certification authority's Certification Practice Statement (CPS), a valid digital signature is the prerequisite for non-repudiation claims.

Therefore, two parties support the engagement in secure and confidential transactions: the mobile operator and the CA. Together, they are responsible for setting up the PKI and providing an (augmented) mobile network infrastructure. On the one hand, the CA is responsible for issuing X.509 certificates that make up the chain of trust between a SmartSignature user, a service provider, and the mobile operator. On the other hand, the mobile operator has the following responsibilities:

1. Distribution of SmartSignature SIM cards
2. Registration of SmartSignature users and forwarding of certification requests to the CA
3. Life-cycle management of SmartSignature
4. Provision of a guarantee that the uploading of a new service provider (service initialization) into SmartSignature has been carried out correctly

The business model during the operation phase is simple. The SmartSignature user and the service provider are subsidized or have to pay for the ability to engage in secure electronic transactions. During the operation phase, the mobile operator recoups the up-front investment and covers the maintenance and mobile phone user support costs by charging for the provision of a secure messaging infrastructure.

SmartSignature in the Setup Phase

The setup phase engages five players: the mobile operator, the SIM card vendor, the CA, the service provider, and the system integrator.

Given that mobile operators are offering their own value-added services to mobile phone users—services such as conventional short messaging, the storage of messages and phone numbers on the SIM card, and the storage of call parameters—they require approximately 16 KB of free memory for a competitive offering in today's GSM market. This means that a SIM card vendor has to offer at least 32-KB memory SIM cards with a math coprocessor for speeding up the RSA encryption/decryption operations.

In addition, SmartSignature has to be implemented in the SIM card operating system programming language. As a consequence, each SIM card vendor has to provide a software implementation of SmartSignature that is uploaded into the SIM cards during the production phase. Once the SIM card vendor receives an order from the mobile operator, a batch of SIM cards is subject to three distinct personalization activities, or phases.

- Phase A comprises the upload of the GSM profile information and all other personalization items. This is the same personalization process as the one for conventional GSM SIM cards.
- Phase B entails the generation of an RSA public/private key pair. The private key and key identifier are stored in the tamper-proof secured file structure of the SIM card. The corresponding public key is issued in an X.509 certificate signed by the card vendor's CA. At the same time, the Mcert of the mobile operator is preloaded into the SIM card.
- Phase C includes the (optional) preloading of any FormPages and SP Mcerts. This enables a service installation without any prior service initialization messages once a mobile phone user has activated a service.

SmartSignature cards and their batch of corresponding X.509 certificates that contain the public keys are forwarded to the mobile operator. As a rule of thumb, the SIM card vendor recoups the upfront investment of building its SmartSignature implementation by allocating an overhead charge to each SmartSignature order. Phase A and B activities are variable costs: They constitute the premium charged for the delivery of a SmartSignature card instead of a conventional SIM card. The premium also includes the increased cost that covers changes in managing the delivery process, i.e., SIM card logistics.

In the setup phase, the CA has to be instructed about the issuing of X.509 certificates that feature SmartTrust-specific profiles. By the same token, the CA has to offer certification services for the mobile operator's subscribers (i.e., SmartSignature users), service providers, and the mobile operator.

All certificates are published in certificate directories. When engaging in a transaction, a service provider simply retrieves the SmartSignature user's certificate for encrypting messages and verifying the digital signature, whereas a SmartSignature user has to trust the mobile operator to deliver an service provider's Mcert to the mobile phone. CAs have an established business model and thus charge per certificate issuance. However, there are certain costs, such as the acceptance of a SmartTrust-specific certificate profile, delivery to and storage in a directory, and management of certificate revocations that can result in a slight premium over conventional certification charges.

Service providers have to install SmartSecurity and SmartCommerce on their premises. They also have to integrate their applications to accept messages from SmartCommerce. As described earlier, applications are sent to SmartSignature by using FormPages and FormData. Each transaction process and service has to be translated into the FormPage and FormData formats. A phone emulator provided by SmartTrust enables the service provider to design and fine-tune a mobile service before the setup of the network components that are needed in the mobile operator's network for transmitting signed and encrypted short messages. The individual service provider thus bears the costs for setting up and designing a mobile service.

The system integrator is responsible for ensuring that the service provider and mobile operator have installed the necessary components for a functioning infrastructure. Availability of the certificate directories and the certificate revocation lists are also a crucial issue. The group of service providers and the mobile operator have to share the costs for integration work.

The setup phase is the most demanding for the mobile operator, who has to install a registration infrastructure that allows the subscribers to exchange their SIM cards and apply for digital certificates.

TrustManager and TrustMapper support both processes. TrustMapper issues a certificate request to the CA based on the SIM vendor's (blank) X.509 certificate and the subscriber's legal identity (as provided by the registering mobile operator's point of sale). TrustManager is the system component that manages the provision of SP Mcerts to SmartSignature. A service provider's Mcert is an n-tuple (i.e., condensed form) of SP's X.509 certificate (as issued by the CA).

The up-front investments for the setup are rolled into the anticipated operating revenue of the SmartTrust infrastructure.

Managing a Large Pilot of SmartSignature

Once the business enablers in the setup and operation phases are in place, commercial deployment of SmartSignature is ready to roll out to customers. This section describes the lessons learned from a specific pilot delivery of SmartSignature that was designed to ensure the security of a mobile stock trading service in Asia.

Pilot Background

A mobile operator in Asia agreed with the local stock exchange and the leading CA to develop and offer an infrastructure for mobile stock-trading services. SmartTrust provided services and the necessary products with a system integrator to provide a turnkey solution project.

The SmartTrust and mobile network infrastructure connect to the automatic order matching and execution system of the stock exchange's mobile channel service. The mobile operator and the CA provide the network infrastructure and PKI for allowing mobile phone users to route trading orders to the stock exchange. The mobile operator and the CA have "developed" an Mcert for secured electronic transactions. The Mcert is an X.509 certificate with a SmartTrust profile. The certificate is backed by a Certification Practice Statement and guarantees recognition by the involved parties if signature verifications have been executed successfully. In this case, the parties agreed to accept the Mcert for authentication, integrity verification, nonrepudiation, and proven confidentiality services.

SmartSignature works on all GSM2+ phones with a SAT. Subscribers who are interested in the mobile stock-trading service can use the mobile operator's point of sales for exchanging their SIM cards and registering for an Mcert.

The Key Participants

One goal of the project was to prove that it is feasible to set up a wireless PKI. SmartTrust and the system integrator provided the following implementation services:

1. Installation of SmartTrust components on site
2. Support in the delivery of the first SmartSignature SIM cards (i.e., augmented SIM card logistics)
3. Support for integration and user acceptance testing
4. Providing product documentation

The players involved and their respective roles are described below.

The *mobile operator* already was a SmartTrust customer and future provider of the mobile stock trading infrastructure. This mobile operator is one of six competitors for the total population of 5.3 million wireless subscribers and has been one of the most innovative in using value-added services to attract new customers. It has expanded to a base of more than 600,000 customers since its launch in 1997. As part of the mobile trust hierarchy, the operator vouches for the integrity and acceptance of valid Mcerts in conjunction with the CA.

The *stock exchange* had implemented a third-generation automatic order and execution matching system in 2000. The introduction of fully electronic connection to the central market has been the driver for opening up the mobile stock-trading channel, where mobile phone users can choose among different brokers for executing their trading.

The Certification Authority is one of the recognized institutions under the electronic transaction regulation mandated by the regional government. Rules and regulations to a large part determine the nature and process of the registration at the mobile operator's point of sale. The government's regulation for electronic transaction has two functions. First, it gives electronic records and DSs the same legal status as that of their paper-based counterparts. Second, it establishes a framework to promote and establish the operation of CAs.

The *certification authority* provides certification services to the mobile operator, the stock exchange, and the mobile phone subscribers (e.g., SmartSignature users).

The *SIM card vendor* provides SmartSignature with a preloaded RSA key and the mobile operator's Mcert.

The *system integrator* is the front-level support for setting up and testing the SmartTrust and other system components.

SmartTrust delivers products, documentation and support for ensuring the delivery of SmartSignature from the SIM card vendor.

Revenue Model

The revenue model for a fully functioning SmartTrust solution is based on a monthly (premium) subscriber fee for SmartSignature users. The fixed fee depends on the amount of short messages sent per month, and there are three levels available. At the moment, there are no plans to charge for the upgrade to a SmartSignature-enabled SIM card. However, there is a fee for issuing the subscriber's certificate (i.e., end user certificate). This fee accrues for the CA.

There is no revenue sharing between the stock exchange and the mobile operator because both parties have agreed to nonexclusive usage. That means that the mobile operator will have to recoup some of the upfront investment costs with the integration of more service providers.

At the moment, no information is available as to whether the stock exchange and the associated brokers will demand premiums for mobile transactions.

Pricing of SmartTrust Components

SmartSignature has been delivered as part of a pilot delivery and, as such, the pricing scheme for SmartTrust system components does not cover all the development costs. In this particular business case, the pricing scheme was based on:

1. A one-time fee for each component
2. An annual maintenance fee (approximately 10—20 percent of the one-time fee)

Some of the component fees are restricted for the anticipated pilot customer group, with the possibility of a reasonably priced upgrading fee. In addition, there was a yearly fee for each active SmartSignature user (approximately 0.5 percent of the purchase price from the SIM card vendor).

Security in a Mobile Trust Hierarchy

Even though SmartSignature is a technical application, as part of pilot planning SmartSignature's developers had to consider security, legal, and macrobusiness processes. SmartSignature's core value proposition centers around providing an infrastructure that allows secure mobile transactions with the highest security standard available. However, systems are complex and therefore have emerging properties that can be discerned only when they are in operation. The desired security level is a matter of precise definition, the technical properties, and the usage behavior of the parties that operate and use SmartSignature.

Moreover, security services are based on contracts that define and assign responsibilities for insurance and control to ensure that an underlying infrastructure is maintained in the appropriate manner. Thus, anybody who wants to build such a complex system faces interesting challenges in aligning the interests of those parties who are setting up the system, those who are maintaining it and guaranteeing its security, and those who are using it. To a large extent, these challenges lie outside the scope of a software engineering company.

For example, legal compliance and insurance contracts became important foundations of the system. Each party had to understand the risks and responsibilities that would be assumed during the setup and operation.

The SmartSignature experience suggests that such a system is driven primarily by concept. Adequate documentation and illustrations about the future roles and responsibilities of interacting players who contribute to the provisioning of a secure mobile transaction infrastructure must be comprehensive and leave no open questions.

Stakeholder analysis and a detailed overview of the contractual obligations help to clarify any gray areas.

Lessons of the Pilot Delivery

The pilot delivery was a prime example of how unanticipated issues can arise. In the pilot delivery, it was very important to preserve the integrity of the concept, because even seemingly minor changes could alter the expected security level. Several legislative issues appeared to hamper the design, until a careful analysis revealed that security requirements were met.

Even though this project centered on software and telecommunication technology, it entailed a lot of political work. Clearly there was a need for different levels of documentation and a careful understanding from the beginning to make appropriate decisions concerning how deeply each party needed to understand the system. Therefore, the level of involvement of each party participating in the project needed to be considered carefully.

One result of the project was the realization that an open PKI system did not seem to be feasible because it was important to have only one CA in the beginning of the project (to avoid complications). It was evident that concentrated business process engineering was required, coupled with a legal design that ensured a clear division of responsibility when the infrastructure was used and maintained.

Customers' expectations needed to be carefully managed, and good communication efforts between the customers and the affected parties were necessary to prevent any difficulties from suddenly appearing once the project was further along.

Importance of the Customer's Experiences

The pilot delivery reinforced the importance of showing the customers the advantages of the SmartSignature concept. It is crucial to stress that the customers are buying an entire concept and a wireless security system, not just one component.

Because the customers' technical needs and the services that the customers intend to use with SmartSignature differ across markets, there was pressure to customize the SmartSignature product offering.

This was a problematic issue because there was always the risk of compromising security procedures with customization. Another vital issue that arose was the need to provide comprehensive support for the SP at all phases of development and setup. This entailed augmenting cross-company communication efforts, maintaining extensive installation and technical support functions, and comprehensive product documentation and training. Therefore, SmartSignature was to a large extent an augmented product offering with a high service content, which required extensive company resources to ensure customer satisfaction.

Implications to the Business Model

From the large-scale pilot delivery, it became clear that delivering a high-security wireless PKI infrastructure with very constrained functions to customers was technically feasible but not commercially viable. Additional functionality and features would have been required to satisfy the mobile operator's and the service provider's customer needs. Another problem was the many players involved—SmartTrust, the mobile operator, the SIM card vendor, the CA, the service provider, and the system integrator—had their own demands and expectations. This required unrealistic amounts of time and effort in communicating and clarifying the positions of all the players. The large number of players in the project also led to the division of the revenue generated by the project between several entities. Therefore, as a lesson from this project, the anticipated revenue model for high-security wireless PKI infrastructures must be realistic and take into account the business and financial requirements of the many different players involved in the process.

One conclusion of the pilot experience was that steps like the SIM card exchange may be too costly for a single-purpose solution with only one service provider, such as using SmartSignature to enable stock trading, and perhaps should be used only when there are a number of features that can be offered to a critical mass of service providers and mobile terminal users.

Implications for SmartTrust Business Strategy

The pilot delivery of SmartSignature in Asia provided some very useful business insights for SmartTrust. The pilot made it evident that an extended mobile PKI infrastructure could meet all the security requirements and end user needs but that it might pose very challenging business issues that would hinder commercial deployment. It became clear that the implementation of an end-to-end mobile trust hierarchy is a complex and expensive process. Even though SmartTrust made a careful review of business processes to match unanticipated requirements for additional customer support during the development and setup phases, the overall pilot required more time, money, and effort than had been anticipated.

In addition, there are implications for the sales and after-sales processes: selling wireless PKI security systems is not a one-time sale of a single component but rather the selling of an entire mobile security system. The number of players involved made it essential to stipulate their respective responsibilities from the very beginning. There also must be a sound legal footing that clearly indicates upcoming liabilities once the wireless security system has been set up. Many of these issues are outside of the scope of a product like SmartSignature.

Based on the experience in the pilot delivery, SmartTrust made significant changes in its SmartSignature product strategy. The new product strategy emphasizes offering a comprehensive wireless security system that provides the infrastructure operator (i.e., the mobile operator) with a value-added service platform that combines SIM device management capabilities and the provision of security services. Putting more emphasis on the role of the mobile operator in generating trust helps to focus the business case and streamline the implementation.

SmartTrust is moving to a value proposition that allows cross-subsidization between SIM card management tools and a launch pad for innovative mobile services. The SmartTrust Wireless Internet Browser supports flexible implementation of e-services for mobile terminals, while the delivery platform and security plug-ins can be used in response to different customer requirements for setting up a versatile, multipurpose wireless security environment.

Next Steps with SmartSignature

The lessons learned from SmartSignature pilot projects have helped SmartTrust to streamline its mobile PKI infrastructure on the business side and build technical enhancements into its latest releases. A 2001 implementation of mobile digital signatures with Vodaphone in the United Kingdom demonstrated this new approach. Vodaphone serves as the trust provider for wireless digital signing of travel and expense reports by staff members at the Department of Trade and Industry in the United Kingdom.

Many of SmartSignature's original mobile trust and security elements are built into this application, but with fewer participants to coordinate, the complexity and expense of the implementation are greatly reduced. This puts the spotlight back on the original driving force behind SmartSignature—delivering a secure mobile transaction in the simplest possible package.

CHAPTER 14

The ETSI Smart Card Platform

There is a convergence of smart card platforms (SCPs) being led by the most widely and intensely used smart card, the SIM. This unified multi-application SCP is being codified by ETSI's SCP project under the chairmanship of Dr. Klaus Vedder. The platform is called the UICC (which does NOT stand for Universal Integrated Circuit Card or anything else). Details of the project, the minutes of its plenary meetings and all of its documents are available at

http://docbox.etsi.org/tech-org/scp/Document/scp

The evolution from the single-function, purpose-built card to a multipurpose application platform while maintaining a maximal degree of backward compatibility is not an easy task and can generate some fairly messy intermediate constructs along the way. As of this writing, the journey is really only starting, but we can already see in the distance some of the defining features of where we are going. This final chapter is about the next step in the evolution of the SIM, the UICC.

The defining property of the UICC is that it carries applications from different providers. For example, it could have a USIM application provided by Vodafone, a credit/debit application provided by Citibank, travel and entertainment applications provided by American Express, and a loyalty application provided by Cathay Airlines. Today all of these applications would be on separate cards and their implementations would assume that they owned the cards and everything on those cards. You can imagine the challenge of getting these headstrong folks to cooperate with each other, but that is exactly the challenge of the SCP.

The first step in the evolution from multiple pieces of plastic to one smart chip with multiple applications is to (gently) lift the bag of bits of each application off its card and put those bags together side by side on the same card and run them one at a time. Each has its own set of data files and each is the only kid on the block when it is active. The next step is to get them to share data. After all, if you change your business telephone number, you don't want to have to give the new number to the same card four times. You want to report it once and have all the applications share it. The final step is to enable the applications to run concurrently just like they do on your laptop. You want to use the USIM application to get on and stay on the network while you use the Citibank application to pay for a music download for which, of course, you want to collect airline points.

Running each application in isolation is not particularly difficult and not particularly useful. You select one application, interact with it, kill it, select another application, interact with it, and so forth. This is the model of the Global Platform. You have fewer pieces of plastic in your wallet or purse but imagine having to reboot Windows each time you wanted to change from reading your e-mail to working on your Power Point presentation, never mind not being able to send an attachement in an e-mail. Nevertheless, multiple applications running concurrently are a very large conceptual leap for the people who have been used to thinking of smart cards in terms of the printing on the outside rather than the bits on the inside. It is a small step for technology but a very big step for brand managers.

When you run one application at a time in isolation, the card acts like a stand-alone card as described in the previous chapters. Therefore, we needn't spend any more time on this first step. What is interesting from a technical point of view is the second and third steps in the evolution of single-application cards to a multi-application platform: data sharing and concurrent execution. These two topics as defined by the SCP are what we cover in this final chapter.

Before diving into the technical details, it must be observed in passing that some applications on the multi-application UICC platform are more equal than others. Even though they share the file system and the processor, some belong to the entity that actually issued and owns the physical card. These issuer applications are responsible for the overall coherency of the card and thus have the ability to keep everybody in line.

If you were paying amazingly close attention in Chapter 9, you would remember that TAR value 00 00 00 was allocated to the Card Manager. This is the capo de capo of applications on the UICC. In the case of a 3GPP UICC with a USIM card, the Card Manager application would be owned by the network operator and one can certainly expect that the operator's applications will have rights and privileges that are not given to other applications (Figure 14-1).

Managed Data Sharing Using Access Control Lists

Perhaps the knottiest architectural problem that the UICC has to solve is managing the rights and privileges of all the entities represented on the card. Who gets to read the data in this file? Who gets to update the data in that file?

Figure 14-1
Architecture of the UICC.

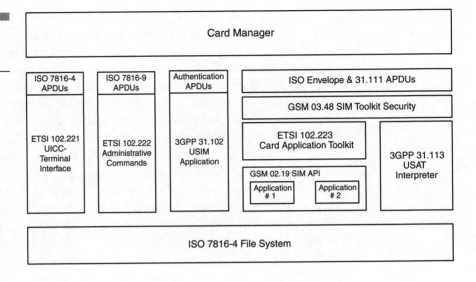

A set of rules for describing who can do what to a file on a UICC card is called an *access control list*, or ACL. Bob Daly and Peter Neumann invented ACLs in 1965 as part of the Multics Project at MIT. The UICC uses the syntax of ISO 7816-8 to describe ACLs for the files in the UICC file system (Figure 14-2).

Associated with every file on the UICC is a set of parameters called the File Control Parameters, or FCP. These parameters describe the size of the file, the name of the file, and, what is of interest here, the ACL for the file.

The things that are listed in an ACL are access rules. Each access rule in the ACL associates a Boolean expression in key references with a file operation. For example, an access rule might be:

```
READ: Key #1 OR (Key #2 AND Key #3)
```

This translates to:

```
This file can be read only if Key #1 has been
successfully used in its authentication protocol or Key
#2 and Key #3 have both been successfully used in their
authentication protocols.
```

Henceforth, the phrase "has been successfully used in its authentication protocol" will be replaced with "has been authenticated" where

The ETSI Smart Card Platform

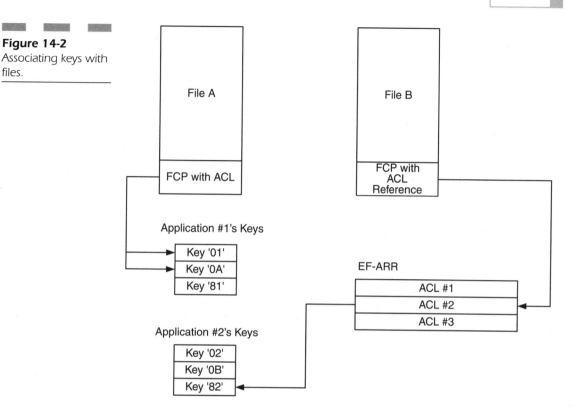

Figure 14-2
Associating keys with files.

the entity with knowledge of the key is being authenticated. We can carry this condensation convention one step further and use the phrase "Key #1 can read the file" as shorthand for "the file can be read if Key #1 has been authenticated."

In ISO 7816 parlance, the file operation is called the *access mode* of the rule, the Boolean expression is called the *security condition* of the rule, the rule itself is called an access rule, and the ACL is called a set of *security attributes*.

Almost all of the access rules associated with files in today's 3GPP USIM application and UICC card contain only one key reference. There are no ORs or ANDs involved. The rules take the forms "Key #1 can READ the file," "Key #2 can WRITE the file," and so forth.

With an eye toward the future capability to represent more complicated rights and privileges, we start to have cards that contain applications from multiple providers. A file containing American Airline AAdvantage points could, for example, have the following ACL:

```
READ: (Cardholder) OR (American Airlines)
WRITE: (Cardholder AND American Airlines) OR (Cardholder
AND WalMart)
```

Strangely missing from the Boolean operators provided by ISO 7816-8 is the NOT operator. This precludes us from writing rules such as "This file may not be read if Key #7 has been authenticated." It also precludes us from representing complicated rules in disjunctive or conjunctive normal form. Perhaps NOT will be added when the need for it finally arises.

Associating Access Control Lists with Files

Before discussing the details of constructing ACLs, we need to describe how ACLs are associated with files. There are only two types: direct and by reference.

The direct way is to put the ACL right into a file's FCP. With this approach, it is easy to find the ACL when you're wondering whether an operation on the file is allowed. If there are only a few files on the UICC or if every file has a unique ACL, then this approach makes sense.

If you look at the files associated with the USIM application, however, you'll notice that there are many files and that there are relatively few uniquely different ACLs associated with them. We could burn up lots of precious space on the UICC by storing these few ACLs again and again in the FCPs of all the files. A better approach for the files associated with the USIM application would be to put the ACLs into a file of their own and to point to this file from the FCPs of the files that are to be protected by them. This is the "by reference" approach.

Of course, nothing says that all the ACLs have to be in one big file. As long as we have the ability to point to the ACL in a file, one large file can contain all the ACLs or many files containing ACLs. Because the ACL files contain their own ACLs, i.e., they can be read and written only by certain keys, which of these two approaches is taken depends completely on the overall security administration policies on the UICC. If the card issuer administers all ACLs, then one big file is most convenient. If each application administers the ACLs associated with its files, then the second approach will be more attractive.

An ACL reference is simply a TLV that consists of a file identifier and a record number:

Tag	Length	File Identifier	Record Index
0x8B	0x03	0x4567	0x04

If you found this TLV in the FCP of a file, then you would go to file 0x4567 and read record 4 to find the ACL for the file. So where is file 0x4567? Look for it in the current directory—the directory containing the file for which you are trying to find the ACL—first and, if it is not there, look for 0x4567 in the directory that is the parent of the current directory. Keep this up until you either find 0x4567 or you top out at the MF or the ADF. With this search strategy, we can store all the ACLs in one or more files in the MF or the application's ADF.

Once we find 0x4567, we read out record 4 and that is the ACL for the file. Obviously, 0x4567 is a record file that can contain records of variable length because, as we will see, ACLs can vary widely in size.

ISO 7816-9 describes an even shorter version of the ACL reference TLV. It is only 3 bytes long and omits the File Identifier:

Tag	Length	Record Index
0x8B	0x01	0x04

To use this form, there has to be an implicit or default file identifier. On the UICC, the default ACL file file identifier is 0x2F06 so this reference TLV is just a shorthand version of the following ACL reference TLV:

Tag	Length	File Identifier	Record Index
0x8B	0x03	0x2F06	0x04

The generic name for 0x2F06 is EF_{ARR} where ARR stands for Access Rule Reference. EF_{ARR} is a variable-length record file. You search for EF_{ARR} the same way you search for any other ACL file. Look first in the current directory for 0x2F06. If it's there, use it. If not, look in the parent directory to the current directory. Keep this up until you find EF_{ARR} or go to the parent of the MF or the application's ADF.

The use of EF_{ARR} centralizes all the access control rules and this eases access rule administration. To build an ACL for a file, you can pick and choose among the available rules or you can add new ones.

If we think of a particular ACL as a security policy, this centralized approach makes it difficult to see what policies are implemented on the card. Each file can select a different set of rules from EF_{ARR}, so to get a list of the policies implemented on the card we have to visit each file. The second approach, in which each policy is put into a different ACL file and then files point to the policy file that governs all access to them, makes maintenance a little more difficult compared with the central database of rules (but not as bad as having rules in every file) but makes the access control policies more visible.

There is a third form of ACL reference TLV that includes an index of a security environment. This form is not often used on the UICC so we won't elaborate here.

Coding Access Control Rules

Now we know how to find the access control rules associated with a file. What do we find when we get there?

There are two ways to code access rules: compact and expanded. Both are TLV templates. The template tag for a compact encoding is 0x8C and the tag for an expanded encoding is 0xAB. We will concentrate on the expanded encoding because it is the most useful and because compact encoding is nothing more than a more efficient way to represent a subset of the rules that can be represented with expanded encoding. The details of both encodings can be found in ISO 7816-9, ETSI TS 102.221, and ETSI TS 102.222.

The value field of the expanded encoding TLV consists of a series of pairs of TLVs:

```
0xAB <length> <access mode> <Boolean expression> ...
                <access mode> <Boolean expression>
```

The file operation in the access mode is allowed only if the Boolean expression immediately following it evaluates to TRUE. Typically, each possible file operation on the file appears in the template at most once. If an operation doesn't appear, then it isn't allowed under any circumstances.

Access Mode TLV

The access mode TLV is of one of two possible types: generic or specific. The generic form starts out with the tag 0x80 and has a value field of only one byte that is a generic description of the operation. Table 14-1 gives the generic access mode TLVs for elementary files.

TABLE 14-1

Access Mode TLVs for EFs

Tag	Length	Value	File Operation
0x80	0x01	0x01	READ
0x80	0x01	0x02	UPDATE
0x80	0x01	0x04	WRITE (deprecated)
0x80	0x01	0x08	DEACTIVATE
0x80	0x01	0x10	ACTIVATE
0x80	0x01	0x20	TERMINATE
0x80	0x01	0x40	DELETE

The WRITE operation is an historical oddity that logically combines new bits with existing bits in the file and isn't supported on the SCP platform. The UPDATE operation is what the rest of the world knows as WRITE.

The value fields can be combined so that an access mode TLV value field of 0x03 would mean that the following Boolean expression applied to the READ and UPDATE operations.

When you use a generic access mode description it is up to the smart card operating system to map the generic description onto the actual commands that the smart card supports. For example, the operating system would know that its UPDATE BINARY and UPDATE RECORD commands, with whatever class and instruction bytes they might have, go with the UPDATE access mode and evaluate the Boolean expression associated with UPDATE when either of these two commands is attempted on a file that contained an ACL with this access mode.

If the generic names for file operations don't give you the degree of control that you need in expressing your security policies, then you can use the specific form of the access mode TLV. In this form of identifying the file operation controlled by the access rule, you give the actual CLA and INS bytes of the command and even the P1 and P2 bytes, if you like. This really gives you minute control!

The specific form starts out with a tag between 0x81 and 0x8F rather than the 0x80 tag of the generic form. The four lower bits of the tag say which bytes of the command (CLA, INS, P1, or P2) are contained in the value field shown in Table 14-2. For example, if the tag were "8C" then the value field would consist of CLA—INS pairs (Table 14-3). The length field is the number of pairs times two. The access mode TLV would be another way to say that the access rule applied to UPDATE operations, namely UPDATE BINARY, UPDATE RECORD, and ERASE BINARY.

TABLE 14-2 Coding of b4...b1 of Specific Access Mode Tag

b4	b3	b2	b1	Meaning
1				CLA byte contained in the value field
	1			INS byte contained in the value field
		1		P1 byte contained in the value field
			1	P2 byte contained the value field

TABLE 14-3 Specific Mode Access Model TLV Example

Tag	Length	CLA1	INS1	CLA2	INS2	CLA3	INS3
8C	06	00	D0	00	DC	00	0E

Key References

The basic unit of the Boolean expression TLV is a key reference. A key reference is TRUE in the evaluation of the Boolean expression if the key is currently authenticated and FALSE otherwise. A key reference also is known as an *authenticatable entity* or a *known principal* in some ACL discussions. In everyday terms, it's very short name for a person such as Sally Green, the cardholder, or an entity such as VoiceStream, the network operator.

The tag "A4" introduces a key reference template. This tag says that this is an authentication TLV. The value field of the key reference template contains additional TLVs describing the key and its associated authentication algorithm. The key reference descriptor TLVs are shown in Table 14-4.

TABLE 14-4 Key Reference Descriptor TLVs

Tag	Key Reference Information
0x80	Authentication algorithm (ISO 9979 and ISO 10116)
0x81	File ID or path to file containing the key
0x82	Directory name of the directory containing the key
0x83	Index of key within the key file (the actual key reference)
0x91	Random number
0x94	Challenge or data item for deriving key
0x95	Usage qualifier

Often the key reference TLV will consist of an 0x83 key index TLV and a 0x95 usage qualifier TLV, with the interpretation being that the index refers to a key of the indicated type in the current application or in the current directory. Table 14-5 shows a key reference TLV of this often-seen variety.

TABLE 14-5 Simple Key Reference TLV

Key Reference Tag	Len	Key Reference	Len	Key Reference Value	Usage Qualifier Tag	Len	Usage (PIN)
A4	06	83	01	01	95	01	08

This key reference template is TRUE only if PIN #1 has been authenticated. It doesn't explicitly say where PIN #1 is stored or which PIN #1 is being referenced in case there is more than one on the card.

The assumption is that the current context on the card—for example, which application is running or what the current directory is—resolves any ambiguity about exactly which Key #1 is being referenced, where it is stored, and whether or not it has been authenticated. More about this later.

Boolean Expressions of Key References

There are two templates used to assemble key references into Boolean expressions of TLVs: an AND template and an OR template. As we mentioned, for some reason there is no NOT TLV, which unfortunately means that you can't always put the Boolean expressions in your access control rules into normal forms.

The AND template starts out with the tag "AF" and the OR template starts out with tag "A0". The value field is then simply what is being AND'd or OR'd together and can of course contain AND and OR TLVs. For example, to simply say "Key #1 and Key #2" you'd use the codes in Table 14-6.

TABLE 14-6 *A Boolean Expression in Key References*

AND Tag	Length	Key Reference Tag	Length	Key Index Tag	Length	Key Reference	Key Reference Tag	Length	Key Index Tag	Length	Key Reference
AF	0A	A4	03	83	01	01	A4	03	83	01	02

An extended format security condition that said "Key #1 OR (Key #2 AND Key #3)" would be coded like this:

```
A0 11 A4 03 83 01 01 AF 0A A4 03 83 01 02 A4 03 83 01 03
```

One of the advantages of this Polish representation of Boolean expressions is that they are programmatically easy to evaluate. Setting aside all of the TLV cruft, here's one way to code it:

```c
#define TRUE   1
#define FALSE  0

#define OR            -1
#define AND           -2
#define END            0

int expression[] = {OR, 1, AND, 2, 3, END, END};
int *bexp = &expression[0];

int beval(), term();

void main()
{
  printf("%s\n\n", beval() ? "TRUE" : "FALSE");
}

int beval()
{
  int e, op;

  op = *bexp++;

  while(*bexp) {
    e = (*bexp < 0) ? beval() : term(*(bexp++));

    if(((op == OR) && e) || ((op == AND) && ~e))
      break;
  }

  while(*(bexp++));

  return e;
}

int v[] = {-1,0,0,1};

int term(keyReference)
{
  return v[keyReference];
}
```

In this little code fragment, term(keyReference) returns TRUE if the key reference has been authenticated and FALSE otherwise. In this example, we hardwired the authentication of key reference 3.

Key Reference Semantics

A key reference is just a name. On some cards such as the Microsoft card, a key reference is associated with the actual name of an entity that can be authenticated like Sally Green or Citibank. In the essentially single-application ISO 7816 series of standards, the semantics of key references are left undefined because who or what is associated with a particular key reference on one card can be totally different from who or what is associated with the same key reference on another card. The association between a key reference and a particular authenticable entity is a matter of coordination between the card and the application using it.

Showing its ancestral roots in telecommunications, the application that the UICC most recently had to coordinate with was the mobile telephone handset. Before the advent of the SAT, the handset had conducted the authentication protocols so it was important that the handset and the SIM agreed on who was being authenticated. Table 14-7 shows the current UICC mapping between key reference values and authenticable entities.

TABLE 14-7

Key Reference Semantics for the UICC

Key Reference Value	Authenticable Entity/ Known Principal
0x01	Primary PIN of Application 1
0x02	Primary PIN of Application 2
0x03	Primary PIN of Application 3
0x04	Primary PIN of Application 4
0x05	Primary PIN of Application 5
0x06	Primary PIN of Application 6
0x07	Primary PIN of Application 7
0x08	Primary PIN of Application 8
0x0A	Administrator 1

continued on next page

TABLE 14-7

Key Reference Semantics for the UICC (continued)

Key Reference Value	Authenticable Entity/ Known Principal
0x0B	Administrator 2
0x0C	Administrator 3
0x0D	Administrator 4
0x0E	Administrator 5
0x11	Universal PIN—PIN shared among all applications
0x81	Secondary PIN of Application 1
0x82	Secondary PIN of Application 2
0x83	Secondary PIN of Application 3
0x84	Secondary PIN of Application 4
0x85	Secondary PIN of Application 5
0x86	Secondary PIN of Application 6
0x87	Secondary PIN of Application 7
0x88	Secondary PIN of Application 8
0x8A	Administrator 6
0x8B	Administrator 7
0x8C	Administrator 8
0x8D	Administrator 9
0x8E	Administrator 10

The PIN key references of an application usually are associated with people and the Administrator key references with computer systems. PIN 1 is for everyday authentication and PIN 2 is for special situations as when PIN 1 gets blocked. PINs are typically 4 or 6 numeric digits or a password of some sort that can be remembered by a human and entered into the card in the clear.

The Administrator of an application is the owner of the application or at least the business entity that is using the application to provide a service to the cardholder. An Administrator key, as its name implies, is used to administer the application. For example, the Administrator key would be used to change the level of service that the application provided to the cardholder. Presumably this would be done only after the cardholder had made arrangements with the application provider to pay for these additional services.

Administrator 1 is usually the card issuer. In the case of a UICC carrying only one USIM application, this would be the network operator. Administrator 2 might be a bank. Administrator keys are authenticated with challenge and response protocols. They are rarely transmitted between the terminal and the card in the clear.

Table 14-7 is not really a satisfactory approach to supporting different authenticable entities for each application on the UICC. First, it provides only three entities per application. This is an unfortunate throwback to the early days of the smart card when there were only three entities on the whole card: CHV1, CHV1, and ADM.

Moreover, associating keys with nameless application indexes—Application 1, Application 2, etc.—and then having to associate application indexes with specific applications introduces an administrative burden and cost that has no return. It will also be a source of error that can easily be avoided. Each application should have its own key reference name space with perhaps a name space shared by all applications for keys such as the Universal PIN.

The operating system on the UICC maintains a list of all the key references that are currently authenticated.

```
AuthenticatedP[key_reference] = TRUE
```

if key_reference is currently authenticated

```
AuthenticatedP[key_reference] = FALSE
```

and if key_reference is not currently authenticated.

When the time comes to evaluate a security condition to see if a certain file operation is allowed, the operating system consults this list to retrieve Boolean values for all the key references in the security condition associated with the file operation.

After associating TRUE (has been successfully authenticated) or FALSE (has not been successfully authenticated) with each key reference in the Boolean expression, the operating system evaluates the entire expression. If the entire expression evaluates to TRUE, the operation is permitted. If the entire expression evaluates to FALSE, the operation is not permitted.

Authentication of Key References

Setting up ACLs is half the problem. Authenticating key references is the other half.

Key references are authenticated using one of the authentication APDUs, VERIFY PIN or EXTERNAL AUTHENTICATE. The terminal—in our case, the handset—sends a VERIFY PIN APDU to the UICC when the subscriber enters a PIN value to activate the UICC as a whole (the Universal PIN) or when a particular application is activated (application PIN). The handset sends an EXTERNAL AUTHENTICATE APDU to the card to authenticate one of the administrators of the data on the UICC such as the network operator. Both APDUs include the key reference they are attempting to authenticate. This is the reason it is critical that the UICC and the handset agree on who is who.

A great mystery about which all the standards are surprisingly quiet is the whereabouts of the data such as the PINs and the passwords that are to be used to figure out if an authentication attempt for a particular key reference succeeds or fails. One reason the standards are silent is that a number of different schemes are used and nobody wants to change them. Another reason is that "security through obscurity" has some lingering cachet in smart-card security circles.

Some UICC cards store the keys in regular data files and use ordinary ISO 7816-4 APDUs such as UPDATE BINARY to write keys into these files. Other UICC cards provide special APDUs for handling keys and store the keys in special places on the card completely outside the file system and thus invisible at the UICC interface. In either case, you can be sure the authentication data is protected by ACLs just like other data and that, as always, the conditions of these ACLs have to be met before the keys can be handled.

What may be different from ordinary data is the authentication protocols that are used to authenticate key references in the ACLs protecting authentication data. For example, a key being written to the card may be encrypted or "wrapped" when it is transmitted to the UICC and the cryptographic wrapping not only protects the confidentiality of the key as it moves from the terminal to the card but it also simultaneously authenticates the entity sending the key.

The cryptographic material for each application is typically associated with the ADF of the application. This information may be in regular files in the ADF or it may be in a special area of the ADF. As we will see below, activation of an application starts with selecting the ADF, and the first thing an application likes to do when it starts is make sure it knows who it is talking to. Said another way, typically you can't do much with an application until you have authenticated yourself to it so that it can access the files it needs to provide you with the service it renders.

Where and how authentication information is actually stored on the card is really only a problem for the card issuer and not somebody such as yourself who is providing applications for the card. You can authenticate your key references using VERIFY PIN and EXTERNAL AUTHENTICATE commands and you can even change your PINs using the CHANGE PIN APDU. You probably will have to work with the card issuer to change your administrator key.

Application Activation and Concurrent Execution

The ACL mechanism described above lets applications on a multi-application SCP card share space; i.e. the data in the file system. These applications also have to share time. The have to execute concurrently and cooperate in delivering services to the cardholder.

Consider delivering multi-media content to a mobile device. This application is on the planning boards of every European telecom operator and on the street in Japan. There will be multiple service providers and multiple trust brokers involved in what appears to the user as a single service. The network operator's application, Vodafone's USIM for example, will have to authorize access to the mobile network. The content provider's application, Vivendi's ISIM for example, will have to authorize use of audio or visual stream. The credit/debit application, Visa's ESIM for example, will have to authorize payment for the use of the bits. And finally, the loyalty application, American Airline's LSIM, will have to authorize accumulation of loyalty points for the transaction.

What must appear to the user as an atomic transaction—"Play Elevator Operator by The Rays"—requires concurrent execution and cooperation among multiple applications on the UICC, each with its own security protocols and its own keys.

There are early signs that the UICC is countenancing concurrent execution of applications. The Java Card virtual machine includes the capability to context switch among applets and the SCP standards include the notion of "first level" applications starting and using "second level" applications. The SCP platform hasn't stepped up to real multi-tasking yet but driven by need it's flirting with the idea.

The Application Directory and Application Activation

Every SCP compliant card has a record file in the root directory called EF_{DIR}, which is the directory of all applications on the card. The file identifier of the application directory is 0x2F00. EF_{DIR} contains one record for each application. Each record consists of a globally unique application identifier and a human-readable string that describes the application all wrapped up in a compound application template TLV.

For example, a record in the EF_{DIR} might be:

Byte Order	Byte Value	Description
1	0x61	Tag — Application Template
2	0x15	Length
3	0x4F	Tag — Application Identifier (AID)
4	0x06	Length
5—10	0xA0 0x00 0x00 0x00 0x99 0x01	Application with the AID 0xA0 0x00 0x00 0x00 0x99 0x01
11	0x050	Tag — Application Label
12	0x0C	Length
13—24	0x42 0x6F 0x6E 0x67 0x6F 0x20 0x4D 0x6F 0x62 0x69 0x6C 0x66	"Bongo Mobile"

An application is activated by sending a SELECT command to the card that contains the application's AID. For example, the SELECT command:

CLA	INS	P1	P2	P3	Data
0x00	0xA4	0x04	0x00	0x06	0xA0 0x00 0x00 0x00 0x99 0x01

would activate the Bongo Mobile application.

The syntax and semantics of AIDs are described fully in ETSI TS 101.220. AIDs can be up to 16 bytes long. The first 5 bytes of an AID identify the entity called the *registered application provider* that built the application. The remaining bytes say which of the applications built by this provider this one is. In the above example 0xA0 0x00 0x00 0x00 0x99 identifies Bongo Incorporated. The trailing 0x01 was assigned by Bongo Inc. to its mobile communications application.

If you are going to be developing applications for the UICC, you should obtain an AID so that your applications can be uniquely identified, registered in the UICC's EF_{DIR} and activated by your customers. AIDs can be obtained from your national standards body. In the U.S. for example that is the American National Standards Institute (ANSI). What you will get is the first five registered application provider that bytes identifier that identify your organization. You are then free to assign bytes 6 through 16 to identify your applications.

The AID of a USIM application starts out with five registered application provider bytes that have been assigned to ETSI: 0xA0 0x00 0x00 0x00 0x09. This is followed by 11 bytes that describe exactly which GSM or 3G network operator owns this USIM. To active a USIM application in order to place a call on the operator's network, the handset sends a SELECT command to the UICC containing all 16 bytes of the USIM's AID in the data field. For example, the SELECT command:

CLA	INS	P1	P2	P3	Data
0x00	0xA4	0x04	0x00	0x10	0xA0 0x00 0x00 0x00 0x09 0x00 0x01 0x03 0xD1 0x00 0x01 0x89 0x00 0x00 0x00 0x01

activates the USIM application of a (hypothetical) Nepalese GSM network operator.

Application Selection

What a terminal device using a UICC typically does when it is turned on is read all the records in the EF_{DIR} to find out what applications are supported on the chip that has been stuck into it. The terminal matches the list of applications that it gets back from the UICC with the list of applications that it knows how to run, a presents a menu of the intersection of these to lists on the screen, and inviting the cardholder to pick the application that he or she wants to run. If the terminal device were an ATM machine, then it would probably ignore all the USIM applications and only offer the credit/debit and account access applications. If the terminal were a mobile phone, it would undoubtedly look for USIM applications.

The point of course is that each application defines its own set of APDU commands and if the terminal doesn't know these commands or know what to do with what comes back when they are sent to the card, then it can't very well run the application effectively. Think of an application AID on a SCP card as a service port on an Internet server. The file transfer service is on port 21. The telnet service is on port 23. The HTTP service is on port 80. Etc. If you make a connection to a port you had better be able to talk its protocol. You can't talk telnet to an FTP port. Similarly, if you start an application on an SCP card you had better be able to talk its APDUs. And also just like an Internet port, not only do you have to be able to talk the protocol, you have to be able to authenticate yourself so you are authorized to do whatever it is you want to do.

The specifications for how you interact with the various applications that might be found on a UICC are published by the people providing those applications. ETSI for example, publishes standards about how to talk to USIM applications. Because ETSI is the nominal provider of all these applications, all USIMs are mandated to behave the same way. There are also specifications for talking to financial applications, travel and entertainment applications and generic identity applications. Generally speaking, cards are way ahead of the terminals when it comes to supporting multiple applications. Mobile phones mostly only know about the USIM application and ATM machines mostly only know about one or maybe two financial applications. The arrival of the Motorola dual-slot Timeport, the Ericsson Electronic Wallet and the Nokia 6310 demonstrate that terminal manufacturers have recognized this problem and the opportunities latent in its solution.

Concurrent Application Execution

In order to start a second application on an SCP card when a first application is already running you use a SELECT command with the second application's AID on a different logical channel. There are four logical channels going into the SCP card and a different application can be running on each one. The low-order two bits on the CLA parameter say which logical channel the SELECT command is being sent on. For example:

CLA	INS	P1	P2	P3	Data
0x03	0xA4	0x04	0x00	0x06	0xA0 0x00 0x00 0x00 0x99 0x01

would start the Bongo Mobile application on logical channel 3 rather than logical channel 0 as in the previous example. All subsequent commands on channel 3 save for a new application-activation SELECT command would go to this executing instance of the Bongo Mobile application. ETSI 102.221 currently says that there can only be one GSM application operating on the card at a time so you can't start Bongo Mobile on channel 3 and then start Big Bend Communications on channel 2 and expect that you can somehow patch these two networks together on your handset.

The fact that there are only four logical channels and therefore the possibility of only four concurrently running applications on the UICC platform is another example of the difficulty of gracefully evolving single-application card constructs to multi-application cards. The four-channel restriction comes from ISO 7816-4. Of course there is nothing stopping one application on the UICC from starting another one and being the second application's proxy to the outside world and this is exactly how UICC application activation is evolving away from ISO 7816's very limited view of application activation.

One possibility is a kind of card inetd, the Unix daemon that starts other services. In a sense this is what the Card Manager of Global Platform is except that it is only focused on getting applications installed on the card and not on activating them and coordinating their behavior. As its name implies, it manages the card on behalf of the card issuer. It doesn't manage the applications.

The ETSI Smart Card Platform

How does the Vodafone application use the Vivendi application for digital rights management? How does the Vivendi application use the Visa application for payment? How does the Visa application use the American Airlines application for loyalty points? All good questions. All currently without answers. There's still lots of work to do on the SCP platform. You are encouraged to track this work on the SCP web site and to participate in the work groups and meetings when you can.

Summary

We have concluded the book with the description of a work in progress, the ETSI Smart Card Platform. This is the evolving SIM. SMS is also evolving. Its 160-byte slow-turnaround packets are evolving to the fat packets of the Multimedia Message Service (MMS) and the rapid-fire packets of GPRS. Both of these changes are part and parcel of the overall evolution of mobile communications from voice medium to data medium.

What we know to date is that just as television wasn't radio with a picture, mobile data isn't the Internet on a handheld. It's something new with its own capabilities and requirements. Working with these requirements will lead to a new style of application.

What we also know is that the most interesting and innovative mobile applications won't necessarily come from today's established mobile telephony players. Visicalc, the killer app for the PC didn't come from IBM or Intel. It came from a couple of guys in a tiny company called Software Arts. Pagemaker, the application that made the Macintosh didn't come from Apple or Motorola. It came from a little company called Aldus.

The killer applications for mobile data are going to come from independent application developers with innovative ideas that somebody forgot to tell it can't be done. They're going to come from application developers like you.

APPENDIX
STANDARDS FOR SMS AND THE SIM

There are lots and lots of standards governing both SMS and the SIM. We list here the ones that have been frequently referred to in this book. All standards including the ones listed here contain a reference section that points to further relevant standards. Following these pointers is a good way to explore the standards landscape.

Third Generation Partnership Project (3GPP)

Latest versions available for free download at: ftp://ftp.3gpp.org/specs/

3GPP Technical Specification Group T (Terminals)—Working Group 2 Mobile Terminal Services and Capabilities

T2 as it is known locally is responsible for messaging in general to the mobile handset. This includes SMS but also the new multi-media messaging service (MMS) and the packet-switches streaming service (PSS). T2 also handles the application-programming interface on the handset called the model for application execution or MExE for short.

3GPP TS 23.039 INTEFACE PROTOCOLS FOR THE CONNECTION OF SHORT MESSAGE SERVICE CENTERS (SMSCS) TO SHORT MESSAGE ENTITIES. Nominally this covers high-speed connections to the SMSCs but it came out after there were lots of proprietary solutions in the field so it hasn't had much success.

3GPP TS 24.011 POINT-TO-POINT (PP) SHORT MESSAGE SERVICE (SMS) SUPPORT ON THE MOBILE RADIO INTERFACE. This is the key document for composing SMS messages.

3GPP TS 27.005 USER OF DATA TERMINAL EQUIPMENT—DATA CIRCUIT TERMINATING EQUIPMENT (DTE-DCE) INTERFACE FOR SHORT MESSAGE SERVICE (SMS) AND CELL BROADCAST (CBS).

3GPP 27.007 AT COMMAND SET FOR 3G USER EQUIPMENT (UE). These two standards define the AT commands to drive your mobile handset from your PC.

3GPP Technical Specification Group T (Terminals)—Working Group 3 Universal Subscriber Identity Module (USIM)

T3 as it is among the standards cognesiti rides herd on the use of the UICC is 3G mobile phones. Many of its core standards have been transferred to ETSI SCP where they are being morphed into general-purpose smart card standards. This frees T3 to focus on the mobile-specific applications such as the USAT Interpreter and annexes to the SCP standards that adapt them for use in 3G mobile networks. As part of this later effort, T3 is responsible for homogenizing all the SIMs in TDMA networks.

3GPP TS 31.102 CHARACTERISTICS OF THE USIM APPLICATION. The USIM application on the UICC is specific to 3G phones and this standard describes this application. It covers the APDUs the USIM application supports in detail as well as the authentication policies and procedures for the USIM.

3GPP 31.113 USAT INTERPRETER BYTE CODES. Defines the nitty-gritty details of creating pages for rendering by the USAT Interpreter.

3GPP TS 42.019 SUBSCRIBER IDENTITY MODULE APPLICATION PROGRAMMING INTERFACE (SIM API); SERVICE DESCRIPTION. Gives the philosophy behind the SAT and its overall architecture in programming language independent terms.

3GPP 43.019 GSM API FOR SIM TOOLKIT. Defines the Java language binding for the SIM Toolkit.

3GPP TS 43.048 SECURITY MECHANISMS FOR SIM TOOLKIT APPLICATION. This is the 3GPP version of what we referred to in the book as ETSI TS 03.48 and is the most current source of information about 03.48 encapsulation and security.

3GPP 51.011 SPECIFICATION OF THE SUBSCRIBER IDENTITY MODULE-MOBILE EQUIPMENT (SIM-ME) INTERFACE. This standard specialized ETSI TS 102.221 for 3G SIMs. It is the direct descendent of the original SIM standard, ETSI TS 11.11.

3GPP TS 51.014 SPECIFICATION OF SUBSCRIBER IDENITY MODULE—MOBILE EQIPMENT (SIM-ME) INTERFACE FOR SIM APPLICATION TOOLKIT. Covers the 3G-specific proactive commands and event downloads.

European Telecommunications Standards Institute (ETSI) Smart Card Project

Latest versions available for free download at: http://docbox.etsi.org/tech-org/scp/Document/scp/

ETSI TS 102.221 SMART CARDS; UICC-TERMINAL INTERFACE; PHYSICAL AND LOGICAL CHARACTERISTICS. This is the basic smart card standard that describes low-level communication with the UICC smart card, the structure of APDUs, the 20 operational UICC APDUs and smart card security. It is the ETSI SCP analog of ISO 7816-4 and ISO 7816-8. Unlike the ISO specifications, ETSI TS 102.221 can actually be implemented.

ETSI TS 102.222 INTEGRATED CIRCUIT CARDS (ICC); ADMINISTRATIVE COMMANDS FOR TELECOMMUNICATIONS APPLICATIONS. APDUs such as CREATE FILE used for doing administration of the UICC platform are defined in this standard. The idea is that these commands are used only by the card issuer and are not typically found in an application or in a handset or a terminal.

ETSI TS 102.223 SMART CARDS; CARD APPLICATION TOOLKIT (CAT). All the generic proactive commands and event downloads are defined in this standard together with the overall protocol for handling two-way interaction with the terminal. There are some proactive commands that are only available on some network technologies—such as USSD on GSM networks—that are covered in network-specific standards that amend and extend ETSI TS 102.222.

International Organization for Standardization (ISO)

Latest versions available for purchase at: http://www.iso.ch

The ISO smart card standards are currently undergoing an exhaustive reorganization and rewriting so that you may want to contact your national standards body to get the latest draft rather than buy old versions from the ISO.

ISO/IEC 7816-4 INFORMATION TECHNOLOGY—IDENTIFICATION CARDS—INTEGRATED CIRCUIT(S) CARDS WITH CONTACTS—PART 4: INTERINDUSTRY COMMANDS FOR INTERCHANGE. This is the Ur smart card standard and is mostly of historical interest. Unlike ETSI TS 102.221 which takes the intersection of everybody's ideas so that the standard is implementable, ISO 786-4 takes the union of everybody's ideas so that everybody can claim standards compliance. The result of course is that ETSI TS 102.221 cards are interoperable and ISO 7816-4 cards are not.

ISO/IEC 7816-8 INFORMATION TECHNOLOGY—IDENTIFICATION CARDS—INTEGRATED CIRCUIT(S) CARDS WITH CONTACTS—PART 8: SECURITY RELATED INTERINDUSTRY COMMANDS. Widely regarded as the most poorly written of the ISO 7816 standards, it covers security environments and certificates. It started down the path of defining the use of smart cards in public key infrastructures but got off on the wrong foot.

ISO/IEC 7816-9 INFORMATION TECHNOLOGY—IDENTIFICATION CARDS—INTEGRATED CIRCUIT(S) CARDS WITH CONTACTS—PART 9: ADDITIONAL INTERINDUSTRY COMMANDS AND SECURITY ATTRIBUTES. Part 9 covers the administrative commands such as CREATE FILE and the details of access control rules.

INDEX

Note: Boldface numbers indicate illustrations; italic *t* indicates a table.

access control list (ACL), 118—119
   Smart Card Platforms (SCPs) and, associating files with, 272—274
   Smart Card Platforms (SCPs) and, managed data sharing using, 269—272
access mode TLV, Smart Card Platforms (SCPs) and, 271, 275—276
ACCESS TECHNOLOGY CHANGE command, 149
account numbers, 110
Across Wireless (*See also* Sonera SmartTrust), 129, 175, 178, 179
ADMinistration duties, SIM, 118
air links, 111
airport logistics application using SMS, 95—105, **100, 101, 102**
alphabets, 13
animation optional feature, 46, 47, 59—60
AnnyWay Information Center, airport logistics application using SMS, 100—105
ANSI 136—510—B
AOL Time Warner, 122
applets, 14
application commands, SIM Application Toolkit API and, 142, 143—145
application design, 7—9, **8**, 16
application port addressing feature, 46, 53—54
Application Programming Interface (API), 14
Application Protocol Data Units (APDUs), 114—115
application server security, 158
application service providers (ASPs), brokered SMS and, 85—86, 89—91
applications as drivers of wireless, 3
ASCII characters and coding, 26—27
Aspects Software, 127
AT commands, 66
   error codes for, 32 command, 32
   handset communication using, 21—25, 30, 32
   RUN AT COMMAND command and, 145
Atraxis airport logistics application using SMS, 95—105, **100, 101, 102**

AU System, 178
authenticable entity, 276
authentication, in Smart Card Platforms (SCPs), 283—284
authentication protocol, 109, **109**
axsControl Departure Control System, airport logistics application using SMS, 95—105, **100, 101, 102**

bandwidth, 17
   brokered SMS and, 80
   Basic, 176
   Basic XHTML, 193
   battery life, 17
   billing, 2, 108
   SmartSignature and, 260—261
   USAT Interpreter and, 206
Bluetooth, 144—145
boolean operators, in Smart Card Platforms (SCPs), 271—272, 278—280
Bos, Jurjen, 176
Branden, Jonas, 178
brokers for SMS, 79—93
   application service providers (ASPs) and, 85—86, 89—91
   bandwidth limitations and, 80
   Code Division Multiple Access (CDMA) support in, 81
   cost of, 82
   definition of, 81
   global, 82
   GSM support in, 81
   m-commerce example using, 87—91
   message flow through, 82, **83**
   personal digital assistant (PDA) support in, 81
   receiving a message using GET, 86—91, **87**
   registering keywords with, 86—87
   security in, 87
   sending a message using POST, 82—86
   SIM hosting and, 91
   standards and, 80

295

Time Division Multiple Access (TDMA) system
  support in, 81
  tracking messages in, 86
  universal coverage problems and, 91
BROWSER TERMINATION command, 149
browsers (*See also* microbrowsers), 14
  LAUNCH BROWSER command, 146
BT Cellnet, 125
business drivers for SMS, 198—200
byte blobs, 26
byte code interpreters, 176—180
byte codes, 175
  USAT Interpreter and, 191—193, 191
  USAT Virtual Machine and, 212, 215—217, 219*t*

C language, 82, 193, 212, 215
  USAT Virtual Machine and, 221, 222—223
  command, 221
C.S0023, 121
cables, for handset, 20
CALL CONNECTED command, 148
CALL CONTROL command, 148, 152
CALL DISCONNECTED command, 148
callback features, in database to SMS integration example, 70
card application toolkit (CAT), 294
card domain (CD), 248—250, 252
Card Manager, for Smart Card Platforms (SCPs), 269
CARD READER STATUS command, 149
CDGRF 43, 120
Cell Broadcast Service (CBS), 12, 292
CELL BROADCAST command, 148
Cellular Digital Messaging Protocol (CDMP), 80
central vs. local storage, 224
certification, SmartSignature and, 241—265
certification authorities, SmartSignature and, 248—253, **253**, 255, 259—260
Certification Practice Statement (CPS), 255, 258
CHANNEL STATUS command, 149
channels, commands for, 147
character sets, 13
  default character encoding for, 7—bit, 33—37
  CICS transactions, 16
  Ciphering Key Identifier (KIc), 162, **163**, 168
  CLOSE CHANNEL command, 147
  Code Division Multiple Access (CDMA)
  brokered SMS support for, 81
  SIM and support, 119—121
colors, 58
COM objects, 16

COM port, 20
Command Header, 158—160
  Sonera SmartTrust Wireless Internet Browser (WIB) and, 185—188
compact HTML (cHTML), 193, 212
compound TLVs, 138
computation on SIMs, 110—111
Computer Access Protocol 2 (CAP II), 80
Computer Interface to Message Distribution (CIMD), 80
concatenated short messages, 46, 51—52
concurrent execution, in Smart Card Platforms (SCPs), 284—289
content and content providers, 199—200
cost, 17
Counter (CNTR) field security, 165, 172
Crossair, airport logistics application using SMS, 95—105, **100**, **101**, **102**
cryptographic checksum (CC), 166
cryptography, 111
cyclic redundancy check (CRC), 168

Daly, Bob, 270
DATA AVAILABLE command, 149, 153—154
Data-Circuit Terminating Equipment (DCE), 12, 292
Data Coding Scheme (DCS) field, 174
  SMS_SUBMIT and, 44, 49—50
Data Download protocol, SIM, 127
Data Encryption Standard (DES), 165
data sharing, using Smart Card Platforms (SCPs), 269—272
Data Terminal Equipment (DTE), 12, 292
Data Terminal Equipment—Data Circuit Terminating Equipment (DTE/DCE), 12, 292
database-to-SMS integration example, 67—78, **68**
  callback features for, 70
  message handling for, 70—71
  response system for, 72—73
  security for, 71
  selected record display in, 74—76
  selection of actions for, 69—70
  sorting out arriving messages to deliver in, 71—72
  state management in, 73—77, **77**,
databases, 16
default character encoding for, 7—bit, 33—37
Derdack Software Engineering (*See* Message Master Developer Suite)
DHMI flag, sounds, 56
DigiCash, 176

Index

digital signature (DS), 166
   in SmartSignature, 241—265
directories and file system of smart card, 112, **112**
DISPLAY PARAMETERS CHANGED command, 149
DISPLAY TEXT command, 143
downloading applications to SIM, 128—130, 133, 148—154, **150**
drivers, for handset, 20, 26
Dual Tone Multifrequency (DTMF), 22

e-mail optional feature, SMS_SUBMIT and, 60—61
eavesdropping, 111
EEPROM, on SIM, 115, 129
EN 300 812, 120
encapsulation of protocols, 9—11, **10**
encryption, 5—7, 54, 108—109, 165
   air/of air link, 111
   ITU TSS CRC—32, 168
   ROT algorithms in, 168
   SmartSignature, 240—265
   wrapping in, 283
enhanced messaging services, SMS_SUBMIT and, 54—56
ENVELOPE command, 127, 149
   Sonera SmartTrust Wireless Internet Browser (WIB) and, 184—188
environment area namespace, USAT Interpreter, 190
Ericsson, 178
Ericsson R310, in airport logistics application using SMS, 103—105
Ericsson, Daniel, 178
error codes, AT commands, 32
ETSI SMG9, 174
ETSI TS 02.19, 15
ETSI TS 03.48, 15, 158—159, 166
ETSI TS 102.221, 14, 274, 293
ETSI TS 102.222, 14, 274, 293
ETSI TS 102.223, 15, 128, 133, 154
ETSI TS 102.233, 294
Europay, 176, 214
European Commission, 174
European Telecommunications Standards Insatiate (ETSI), 177—178, 293—294
EVENT DOWNLOAD command, 145
event downloads, 148—154, 150
event lists, 144—145
Everett, David, 176
executable programs, 14—15
External Machine Interface (EMI), 80

FETCH command, 136
file system of smart card, 111—112, **112**,
FilmWeb, 179
fire-and-forget model of interfaces, 14
Forms, SmartSignature, 243—244, 257
Forth, 176

GAIT-H-1-1-2-0, 128
General Packet Radio Services (GPRS), 6, 92
general purpose communication commands, SIM Application Toolkit, 142, 146—147
generations of SIMs, 115—118, **116**, **117**
GET CHANNEL STATUS command, 147
GET INKEY command, 143
GET INPUT command, 143
GET READER STATUS command, 146
GET, receiving a message via broker and, 86—91, **87**
Gismo, USAT Interpreter and, 200—204, **201**
Global Open Platform, Smart Card Platforms (SCPs) and, 269
Global System for Mobile Communications (GSM), 2, 3, 24—25, 108
   brokered SMS and support for, 81
   SIM and, 119, 121, 126
   SmartSignature, 240—265
   USAT Virtual Machine and, 221, 222t
Gordons, Edouard, 176
GPA Technology, 66
Grimonprez, Georges, 176
GSM Association, 4, 55
GSM 11.11 standard, 120
GSM 11.14 standard, 128
Guilfoyle, Tony, 176

Hamlin, Colin, 125—127
handoffs, 3
handsets, 48
   AT commands for, 21—24, 25, 30, 32
   communicating with, 21—24
   connection to, 20—21, **21**
   drivers and, 26
   Hello Mobile World application for, 25—37
   HyperTerminal testing of, 20—21, 23—24
   master slave relationship with SIM and, 125
   message status codes for, 30
   network communication using, 24—25
   number of messages sent using, per month, 2
   opening a serial port connection to, 25—26
   pop-up messages for, 31
   ports for, 20

reactive interface SIM and, **134**
receiving messages at, 30—32
testing connection to, 20—21
Harvard architecture, 218
headers
  SMS_DELIVER and, 42, 61
  SMS_SUBMIT and, 42, 43, 61
hex codes, 26—27
  default character encoding for, 7—bit, 33—37
home location register (HLR), 147
hosting, SIM, 91
HTML, 178
HTTP, 9, 178
  Sonera SmartTrust Wireless Internet Browser (WIB) and, 181—188, **181**
Hubwatch, airport logistics application using SMS, 95—105, **100, 101, 102**
HyperTerminal, testing handset connection using, 20—21, 23—24,

IBM, 6
icons (*See also* animation optional feature), 47
iDen phones, SIM support in, 119
IDLE SCREEN AVAILABLE command, 149
iMelody, 57—58, 144
IMSAI, 218
Infrared Data Association, 57—58
Integrated Circuit Cards (ICC), 13, 293, 294
integrated portal, USAT Interpreter and, 202—203, **203**
integration of SMS, 65—78
  database connection example of, 67—78, **68**
  Kannel Open Source WAP and SMS Gateway for, 66
  Message Master Developer Suite for, 66
  message queues in, 66—67, **67**
  Nokia PC Connectivity Software Developers Kit (SDK) for, 66
  SMS Gateway for, 66
  SMS-IT for, 66
Intelligent Short Message Service Center (ISMSC), 80
interfaces, 14, **15**, 129
International Organization for Standardization (ISO), 294
Internet applications (*See also* application port addressing), 46
Internet e mail optional feature, SMS_SUBMIT and, 60—61
Internet Protocol (IP), 122
Internet Service Providers (ISPs), 8

interpreters, 175, 176—180
interworking applications, 47
inward-looking API for SIM, 132—133, **132**
IP Multimedia Service, 122
IrDA ports, 20
IrDA, 144—145, **144**
IS 820, 120
IS—820, 121
ISIM, 122
ISO 7816—4, 127, 149
ISO 7816—9, 274
ISO 8859, 33
ISO/IEC 7816—4, 13
ISO/IEC 7816—8, 13
ISO/IEC 7816—9, 13
ISO/IEC 7816—4, 294
ISO/IEC 7816—8, 294
ISO/IEC 7816—9, 294
ITU TSS CRC—32, 168

Java, 5, 82, 178, 193, 212, 293
  SIM and programming, 128
Java Card, 178, 214, 215—217, **217**, 235, **235**
Java Script, 179
JScript, 82

Kannel Open Source WAP and SMS Gateway, 66
Key Identifier (KID), 162, **163**
Keycorp, 176
keys, for encryption, 108—110
  SmartSignature and, 244, 257
keywords, brokered SMS and, 86—87
"killer applications," 16
known principal, 276

LANGUAGE NOTIFICATION command, 148
LANGUAGE SELECTION command, 149
LAUNCH BROWSER command, 146
length of SMS messages, 4—5
Lille University, 176
line drawings, 58—59
Lisp, 176
LOCATION STATUS command, 148
long messages (*See* concatenated short messages)
Lundh, Per, 178

m-commerce
  example using a broker, 87—91
  SmartSignature, 239—265
  USAT Interpreter and, 204—205
markup languages, 129, 180, 193, 212

Index

master slave relationship with handset and SIM, 125
math extension, USAT Virtual Machine and, 218, **218**
Mcerts, SmartSignature and, 249—250, 257, 258
memory capacity of SIMs, 6, 115, 129
  SmartSignature requirements and, 257
menu design
  SETUP MENU command for, 143
  SmartSignature, 244—245, **245**
MENU SELECTION command, 148, 152
Message Master Developer Suite, 66
message queues, integration of SMS and, 66—67, 67
message status, 30
message types in SMS, 40—41, **41**
message waiting indicators (*See also* special SMS message indication), 46
MExE, 291
Micro—Browser, SIM, 4
microbrowsers, 14, 129—130, 178, 179, 198—200
  Sonera SmartTrust Wireless Internet Browser (WIB), 180—188, **181**
  USAT Interpreter, 188—195, **189**, **190**, 197—210
  USAT Virtual Machine and, 215
Microsoft, 176
mobile banking, USAT Interpreter and, 204—205, **205**
mobile phones, 6—7
modems, 12
  for handset, 20
MORE TIME command, 147
Motorola Timeports, in airport logistics application using SMS, 103—105
MTCALL command, 148
Multics Project, 270
Multimedia Messaging Service (MMS), 291
  SMS_SUBMIT and, 54—55
Multos virtual machine, 176, 214

n-tuples, 249, 258
name spaces
  USAT Interpreter, 189—190
  USAT Virtual Machine and, 212
NATELsicap, 174, 175
network design and SMS/SIM, 7—9, **8**
network identity (NID), 248—250
networks, communicating with, 24—25
Neumann, Peter, 270
next generation wireless, 3
nibble swapping, 62—63

nibbles, 42
Nokia PC Connectivity Software Developers Kit (SDK), 66
non-repudiation, SmartSignature and, 255
NovelSoft, 86
number of messages sent using SMS, per month, 2
numbering plans, 42

objects, 16
OPEN CHANNEL command, 147
Open Interface Specification (OIS), 80
outward-looking API for SIM, 132—133, **132**
over-the-air (OTA) security, 158

packet switches streaming service (PSS), 291
Padding Counter (PCNTR), 165
page string area namespace, USAT Interpreter, 190
pairing a sent message with its response, 170—172
Paradinas, Pierre, 176
Pascal, 176
Pedersen, Thomas B., 199—210
PERFORM CARD APDU command, 146
Perl, 82
permanent area namespace, USAT Interpreter, 190
personal digital assistants (PDAs), 2
  brokered SMS and support for, 81
  personal identification number (PIN), 71, 118
  SmartSignature and, 244
personal identification number 2 (PIN2), 118
picture optional feature, 47, 58—59
ping, 47
PKI security, using SmartSignature, 241—265
PLAY TONE command, 138—140, 144
point-to-point (PP) support, 13, 292
POLL INTERVAL command, 136, 147
POLLING OFF command, 147
pop-up messages, 31
ports, handset, 20
POST, sending a message via SMS broker and, 82—86
POWER ON CARD/POWER OFF CARD command, 146
proactive command protocol, SIM, 125, **126**, 133—142, **137**, 175
proof-of-receipt (PoR), 168—170
Protocol Identifier field, SMS_SUBMIT and, 44, 47—49, 48t
protocol stacks, 9—11

Index

PROVIDE LOCAL INFORMATION command, 144, 145
public key certificates, in SmartSignature, 241—265

Q.1741, 120
queues (*See* message queues)

reactive interface between handset and SIM, 134
real time communications, 92—93, **93**
Real Time Travel example using USAT Virtual Machine, 224—234
RECEIVE DATA command, 147
receiving a message (*See also* SMS_DELIVER), 41
redundancy check (RC), 166, 168
REFRESH command, 148
registration, in brokered SMS, 86—87
remote procedure call (RPC) using USAT Interpreter, 193—195
Reply Path field, SMS_SUBMIT and, 44
Response Packet and Header, security and, 169—170
Rev-C 136—037, 128
roaming, 2, 3, 108—109,
root domain (RD), 248—250, 252
ROT algorithms, security, 168
routing through wireless communication network, 9—11, **10**
RSA keys, in SmartSignature, 255
RUN AT COMMAND command, 145
runtime checking, in USAT Virtual Machine, 217
Runtime Environment (RTE), for USAT Virtual Machine, 216

scheduling, airport logistics application using SMS, 95—105, **100, 101, 102**
Schlumberger, 176, 178
SCP 102.221, 121
secret keys (*See* keys, for encryption)
security, 6, 15, 92, 111, 129, 157—172, 293, 294
  application server and, 158
  brokered SMS and, 87
  Ciphering Key Identifier (KIc) in, 162, **163**, 168
  Command Header in, 158—160
  Counter (CNTR) field in, 165, 172
  cryptographic checksum (CC) in, 166
  cyclic redundancy check (CRC) in, 168
  Data Encryption Standard (DES) and, 165
  in database to SMS integration example, 71
  digital signature (DS) in, 166

ETSI TS 03.48 standard for, 158—159, 166
Key Identifier (KID) in, 162, **163**, 162
over-the-air (OTA), 158
Padding Counter (PCNTR) in, 165
in pairing a sent message with its response, 170—172
proof of receipt (PoR) in, 168—170
redundancy check (RC) in, 166, 168
Response Packet and Header in, 169—170
ROT algorithms in, 168
secured SMS message example of, 166—172
Security Parameter Indicator (SPI) and, 161—162, **161, 162**
SIM Toolkit Security in, 158—159, **159**
SmartSignature, 239—265
SMS_SUBMIT and, 54
  Toolkit Application Reference (TAR) in, 164
  USAT Interpreter and, 205
  User Data Header and, 170
security attributes, 271
security condition, 271
Security Parameter Indicator (SPI), 161—162, **161, 162**
SELECT ITEM command, 144
Sellin, Lars—Erik, 178
SEND, 144
SEND DATA command, 147
SEND DTMF, 144
SEND SHORT MESSAGE command, 144
SEND SS command, 147
SEND USSD command, 147
sending a message (*See also* SMS_SUBMIT), 41
serial ports, 20
SET EVENT LIST command, 151
SET UP EVENT LIST command, 144—145, 148
SET UP IDLE MODE TEXT command, 145
Setec virtual machine, 178
SETUP CALL command, 144
SETUP MENU command, 143
Short Message Centers (SMSCs), 12
Short Message Client Interface (SMCI), 80
Short Message Entities (SMEs), 12
Short Message Peer to Peer (SMPP), 80
Short Message Service (SMS), 2, 4—7, 174—175
Short Message Service Centers (SMCSs), 24—25, 29, 54
Sicap messages, 174
signing PIN, 242, 244
SIM (*See* Subscriber Identity Module)
SIM Application Programming Interface (SIM API), 292

Index

SIM Application Toolkit (SAT), 4, 5, 122—127, 174—175, 198
  application commands for, 142, 143—145
  CALL CONTROL command for, 152
  command details for, 142—148
  DATA AVAILABLE command for, 153—154
  ENVELOPE command and, 149
  event download commands for, 133, 148—154, **150**
  FETCH command for, 136
  general purpose communication commands for, 142, 146—147
  inward and outward APIs for, 132—133, **132**
  MENU SELECTION command for, 152
  PLAY TONE command in, 138—140
  POLL INTERVAL command for, 136
  proactive command flow in, 133—142, **137**
  reactive interface between handset and SIM, **134**
  SET EVENT LIST command for, 151
  SET UP EVENT LIST command, 148
  smart card commands for, 142, 146
  SMS CONTROL command for, 152
  SMS PP command for, 151—152
  STATUS command for, 136
  summary of proactive command action in, 141, 142
  system commands for, 143, 147—148
  Tag-Length-Values (TLVs) in, 137—140
  tags in, 138
  TERMINAL RESPONSE command for, 136, 140—141
  TIMER EXPIRATION command for, 152—153
  TIMER MANAGEMENT command for, 152
SIM Application Toolkit API (SAT API), 127—128, 131—155, **132**, 175
  USAT Virtual Machine and, 212
SIM hosting, brokered SMS and, 91
SIM Micro—Browser, 4
SIM Toolkit Security, 54, 158—159, **159**
SIMAlliance, 179m 189
smart card commands, in SIM Application Toolkit API, 142, 146
Smart Card for Windows, 214—216, **215**, 176
Smart Card Manager, 199
Smart Card Platforms (SCPs), 267—289
  access control lists and
    associating files with, 272—274
    managed data sharing using, 269—272
    access control rules for, coding of, 274—275
    access mode TLV for, 271, 275—276
  application activation in, 284—289
  associating keys with files in, 270—271, **271**
  authenticable entity or known principal in, 276
  boolean expressions of key references in, 278—280
  Boolean operators in, 271—272, 278—280
  Card Manager and, 269
  concurrent execution in, 284—289
  Global Open Platform and, 269
  key reference authentication, 283—284
  key reference semantics for, 280—282
  key references for, 276—278
  security attributes in, 271
  security condition in, 271
  TAR value for, 269
  UICC and, 268, **270**
  USIM and, 269, 271
Smart Card Project, 293—294
smart cards, 5, 13, 14, 15, 110, 111—115, 293—294
  access control list (ACL) and, 118—119
  Application Protocol Data Units (APDUs) for, 114—115
  byte code interpreters and, 176—180
  directories and file system in, 112, **112**
  file system within, 111—112, **112**
  language preference instruction, example using, 112—114
  readers for, 113
SmartCommerce, 257
SmartSecurity, 257
SmartSignature, 239—265
  business enablers of, 253—258
  business model for, 263
  card domain (CD), 248—250, 252
  certification authorities and, 253, **253**, 255, 259—260
  Certification Practice Statement (CPS) and, 255, 258
  certification using, 248—253
  changing service providers and, 245—248
  components of, 254, **254**
  customer satisfaction with, 262—263
  Forms, 257
  forms and templates for, 243—244
  future of, 264—265
  implications of use of, 263—264
  keys in, 244, 257
  Mcert and, 249—250, 257, 258
  memory requirements, 257
  menu design for, 244—245, **245**

network identity (NID), 248—250
non-repudiation in, 255
operation of, 254—256
personal identification numbers (PINs) in, 244
pilot operation using, 258—264
PKI and, 248, 258, 263
pricing of components in, 260—261
revenue model using, 260
root domain (RD), 248—250, 252
RSA keys for, 255
sample use of, 242—243
service selection using, 245—248
setup phase for, 256—258
signing PIN for, 242, 244
SIM card vendor for, 260
SmartCommerce and, 257
SmartSecurity and, 257
system integrator for, 260
trust entities for, 249
trust hierarchy in, 261
trust relationship establishment in, 251—252, **251, 252**
TrustManager for, 258
TrustMapper for, 258
X.509 certificates and, 249—250, **250**, 255, 257—258
SmartTrust (*See also* SmartSignature), 173, 189, 241
SMG9 Standards Committee, 126
SMPP Forum, 80
SMS CONTROL command, 148, 152
SMS Gateway, 66
SMS_COMMAND, 40
SMS_DELIVER, 30—32, 40, 61—63, **61**, 166, 170
  header for, 42, 61
  in pairing a sent message with its response, 171—172
  nibble swapping in, 62—63
  non-address bytes in, 61
  Sonera SmartTrust Wireless Internet Browser (WIB) and, 185—188
  subfields and SMS flags in, 62
  UDH Source Indicator for, 63
  User Data field in, 63
SMS_DELIVER_REPORT, 40
SMS_PP, 148, 151—152, 158
  Sonera SmartTrust Wireless Internet Browser (WIB) and, 184—188, 184
SMS_STATUS REPORT, 40
SMS_SUBMIT, 26, 29, 30, 31, 40, 42—61, 43, 144, 166, 168, 170
  animation optional feature in, 46, 59—60
  application port addressing feature in, 46, 53—54
  concatenated short messages in, 46, 51—52
  Data Coding Scheme field in, 44, 49—50
  enhanced messaging services and, 54—56
  fields in, **45**
  header for, 42, 43, 61
  in pairing a sent message with its response, 170—172
  Internet e-mail optional feature in, 60—61
  Multimedia Messaging Service (MMS) and, 54—55
  non-address bytes in, 43 command, 43
  optional features in, 45t, 46—47
  picture optional feature in, 58—59
  Protocol Identifier field for, 44, 47—49, 48t
  Reply Path field in, 44
  SIM Toolkit security feature in, 54
  Sonera SmartTrust Wireless Internet Browser (WIB) and, 187—188
  sound optional feature in, 56—58
  special SMS message indicator feature in, 46, 52—53
  subfields of, 43—44, 43t
  tags in, 45—46, 46t
  User Data field for, 61
  User Data Header Indicator field in, 44
  USIM Toolkit security, 46
SMS_SUBMIT_REPORT, 40
SMS-IT, 66
SMS2000, 80
software development kits (SDKs), USAT Virtual Machine and, 215—216
Sonera SmartTrust (*See also* Across Wireless), 129, 179
Sonera SmartTrust Wireless Internet Browser (WIB), 180—188, **181**
Sonofon, 198—210
Sony, 122
sound optional feature, 47, 56—58
SP 4027—030,033,034, 120
special SMS message indicator feature, 46, 52—53
spectrum, in wireless communications, 9
SQL, 16
standards development organizations (SDOs), 120
standards, 11—15, **12**, 80
STATUS command, 136
status information, 47—48
storage capacity, 6, 115, 129
storage, central vs. local, 224
subroutine calls, using USAT Virtual Machine, 212
Subscriber Identity Module (SIM), 2, 3, 4, 107—130, 198

Index

access control list (ACL) and, 118—119
ADMinistration duties in, 118
applications on, 174—175
applications placed on, acceptance for, 122, 125
authentication of subscriber using, 118—119
Code Division Multiple Access (CDMA) support for, 119—121
computational ability of, 110—111
Data Download protocol for, 127
downloading applications to, 128—130
ENVELOPE command and, 127
evolution and generations of, 115—118, **116**, **117**
GSM and, 119, 121, 126
iDen phone support for, 119
interface with, 129
ISIM and, 122
Java and, 128
Java Card, 235, **235**
markup languages and, 129
master slave relationship with handset and, 125
memory capacity of, 115, 129
microbrowser and, 129—130
personal identification number (PIN) and, 118
personal identification number 2 (PIN2) and, 118
proactive command protocol for, 125, **126**
Protocol Identifier and, 49
reactive interface between handset and, **134**
security and, 129
SIM Application Toolkit and, 122—127
SIM Application Toolkit Application Programming Interface (SAT API) for, 127—128
smart cards and, 110, 111—115
standards for, 119—122
storage capacity of, 115, 129
Time Division Multiple Access (TDMA) support for, 119—121
UICC and, 121—122, **123**
USAT Interpreter for, 128—130
USIM and, 122, **125**
Subscriber Identity Module Application Programming Interface (SIM API), 15
Subscriber Identity Module Mobile Equipment (SIM-ME), 293
Supplemental Service (SS), 147
Swissair, airport logistics application using SMS, 95—105, **100**, **101**, **102**
Swisscom, 127, 175
   airport logistics application using SMS, 95—105, **100**, **101**, **102**

system commands, SIM Application Toolkit API and, 143, 147—148

T63-31.102, 120
T64-C.S0023, 120
Tag-Length-Values (TLVs), 137—140
   Sonera SmartTrust Wireless Internet Browser (WIB) and, 182—188
tags, 138
TCP/IP, 9
telematic interworking, 47
telephone numbers, mobile, 42
TELEPOINT standard, 120
Telnor, 179
templates, SmartSignature, 243—244
temporary area namespace, USAT Interpreter, 190
TERMINAL RESPONSE command, 136, 140—141, 143, 145, 146
text formatting optional feature, **47**
Third Generation Partnership (3GPP) (*See also* 3GPP*xx* standards), 291—294
Thorsstensom, Tommy, 178
31.101, 121
3G 31.111, 128
3G wireless, 3
3GPP T3, 121
3GPP TS 22.112, 188
3GPP TS 23.003, 13
3GPP TS 23.038, 13, 33
3GPP TS 23.039, 12, 291
3GPP TS 23.040, 13, 24—25, 30
3GPP TS 24.011, 13, 24—25, 42, 292
3GPP TS 27.005, 21, 292
3GPP TS 27.0056, 12
3GPP TS 27.007, 12, 21, 292
3GPP TS 29.002, 42
3GPP TS 31.102, 292
3GPP TS 31.112, 188
3GPP TS 31.113, 14, 189, 292
3GPP TS 31.114, 189
3GPP TS 42.019, 292, 293
3GPP TS 43.048, 293
3GPP TS 51.011, 293
3GPP TS 51.014, 293
3GPP USAT Interpreter (*See* USAT Interpreter), 173
31.101, 120
time, nibble swapping in, 62—63
Time Division Multiple Access (TDMA)
   brokered SMS and support for, 81

SIM and support, 119—121, **119**
timeouts, 147
TIMER EXPIRATION command, 148, 152—153
TIMER MANAGEMENT command, 145, 152
Toolkit Application Reference (TAR), 164
   Smart Card Platforms (SCPs) and, 269
   Sonera SmartTrust Wireless Internet Browser (WIB) and, 180—188, **181**
Transfer Protocol Data Unit (TPDU), 26, 29, 31
Transmission Control Protocol (TCP), 53, 164
transmission time, 17
trust entities, in SmartSignature, **249**
trust relationships, in SmartSignature, 251—252, **251**, **252**
trusted third-party certificates, 241
trusted transactions, 6
TrustManager, 258
TrustMapper, 258
2.5G wireless, 3

UATK, 128
UDH Source Indicator, SMS_DELIVER and, 63
UICC, 14, 293
   SIM and, 121—122, **123**
   Smart Card Platforms (SCPs) and, 268, **270**
Uniform Resource Locator (URL), 193
   Sonera SmartTrust Wireless Internet Browser (WIB) and, 180—188, **181**
USAT Interpreter and, 193
Universal Computer Protocol (UCP), 80
Universal Subscriber Identity Module (USIM), 122, **125**, 292
   Smart Card Platforms (SCPs) and, 269, 271, 272
   Toolkit security, SMS_SUBMIT and, 46
Unstructured Supplemental Service (USSD), 147
USAT, 292
USAT Interpreter, 5, 14, 128—130, 173, 179, 188—195, **189**, **190**, 197—210
   benefits of, 209—210
   billing in, 206
   business drivers for, 198—200
   byte codes for, 191—193
   cost of, 206
   Gismo and, 200—204, **201**
   implementation challenges and strategies for, 207—209
   integrated portal for, 202—203, **203**
   lessons learned in, 210
   m-commerce using, 204—205
   mobile banking using, 204—205, **205**
   remote procedure call (RPC) using, 193—195
   sample page in, 192
   security and, 205
   service selection in, 206—207
   SMS and, 200—202
   subscribers of, 205—207
   Toolkit Application Reference (TAR) in, 164
   USAT Virtual Machine and, 212, 213 command, 212
   user point of view of, 205—207
   Wireless Application Protocol (WAP) and, 202—203, **203**, 210
USAT Virtual Machine, 211—237
   architectures of, 216—217
   arithmetic extension byte codes for, 220
   byte codes and, 212, 215—217, 219t
   C language support for, 221, 222—223t
   central vs. local storage in, 224
   GSM specific system calls in, 221, 222t
   Java Card, 214, 215—217, **217**, 235, **235**
   math extension for, 218, **218**
   microbrowsers and, 215
   Microsoft USAT Interpreter, 218—223
   Multos virtual machine, 214
   name spaces and, 212
   permanent vs. transient pages in, 213
   program installation on, 235—236
   Real Time Travel example using, 224—234
   runtime checking in, 217
   Runtime Environment (RTE) and, 216
   SAT API and, 212
   Smart Card for Windows, 214—216, **215**
   software development kits (SDKs) for, 215—216
   standards for, 215
   storage of, 213
   subroutine calls in, 212
   SYS byte code, 221
   USAT Interpreter and, 212, 213t
   utility functions for, 223, 223t
   variants of, 214—216, **215**
   Visual Basic and, 221
   wild-card byte codes, 221
   Wireless Internet Gateway (WIG) and, 212
   Wireless Markup Language (WML) and, 212, 213, **214**
USB ports, 20
USER ACTIVITY command, 148
User Data field
   SMS_DELIVER and, 63
   SMS_SUBMIT and, 61
User Data Header
   security and, 170

Index

SMS_SUBMIT and, 44
Sonera SmartTrust Wireless Internet Browser (WIB) and, 185—188
User Datagram Protocol (UDP), 53
User Equipment (UE), 12, 292

value-added services, 198—200
van der Hoek, Jelte, 176
VBScript, 82
   sending a brokered message using, 84—85
Vedder, Klaus, 126—127, 268
very large instruction word (VLIW), 31
virtual machines, 5, 175
Visual Basic, 82, 193, 212, 215, 221
Visual C++, sending a brokered message using, 83—84
Vivendi, 122
Vodafone, 128—129
voice calls, 48
VoiceStream, 29, 63

WAP browsers, 48

WAP—260—WIM—20010712—a, 120
Watts, David, 176
Wireless Application Protocol (WAP), 3—4, 178, 179, 189, 192, 200, 210
   USAT Interpreter and, 202—203, **203**
Wireless Internet Browser (WIB), 180—188, **181**
Wireless Internet Gateway (WIG), 178, 179
   Sonera SmartTrust Wireless Internet Browser (WIB) and, 180—188, **181**, 180
   USAT Virtual Machine and, 212
Wireless Markup Language (WML), 180, 193, 212, 213, **214**
Woodsend, Kristian, 125—127, 175
Working Group 3, 3GPP, 292
wrapping, in encryption, 283

X.509 certificates, in SmartSignature systems, 241, 249—250, **250**, 255, 258
XHTML, 193, 212

Zurich Airport and airport logistics application using SMS, 95—105, 100, 101, 102

ABOUT THE AUTHORS

SCOTT B. GUTHERY is one of the best-known names in smart cards. Currently Chief Technology Officer of Mobile-Mind, a provider of wireless system software and trusted applications, he was the lead designer of Microsoft's Smart Card for Windows. He also led the team that developed the first Java Card. A regular contributor to GSM and 3G standards, he holds four patents for smart card applications and real-time systems. He is the convener of the Architecture Working Group of the European Telecommunications Standards Institute (ETSI) Smart Card Platform (SCP) project.

MARY J. CRONIN, Mobile-Mind President, is the author of five books on technology and business strategy, including *Doing Business on the Internet*, a groundbreaking work that has been translated into 10 languages. As a Professor of Management at Boston College, Dr. Cronin specializes in Electronic Commerce, International Telecommunications, and Wireless Information Management.